# THE
# ASTONISHED
# TRAVELER

# THE
# ASTONISHED
# TRAVELER

*William Darby, Frontier Geographer*
*and Man of Letters*

*J. Gerald Kennedy*

*Louisiana State University Press*
*Baton Rouge and London*

*Copyright © 1981 by Louisiana State University Press*
*All rights reserved*
*Manufactured in the United States of America*

*Designer*: Joanna Hill
*Typeface*: Bembo
*Typesetter*: G&S Typesetters, Inc.
*Printer and binder*: Thomson-Shore, Inc.

*Published with the assistance of the*
*Department of Geography and Anthro-*
*pology, Louisiana State University.*

*Library of Congress Cataloging in Publication Data*

Kennedy, J. Gerald.
The astonished traveler.
Bibliography, p.
Includes index.
1. Darby, William, 1775–1854.  2. Geographers—
United States—Biography.  I. Title.
G69.D35K46     917.3 [B]     81-3711
ISBN 0-8071-0886-3     AACR2

*To my dear Mary Ann*

# Contents

# Maps

# Preface and Acknowledgments

This book represents the culmination of a detective story, a mystery of concealed identity, which came to light in 1973. That year I began research on the American magazine tales of the 1830s, intending a preliminary study of the relationship between the short story genre and the literary periodicals of that day. Surveying the *Casket* (a Philadelphia monthly), I discovered a long series of tales, mainly Cooperesque border narratives, by a writer signing himself "Mark Bancroft." The standard literary histories and reference works offered no information about this prolific author, and my essay (which appeared in *American Transcendental Quarterly* in 1974) consequently made only passing mention of the Bancroft tales. But the puzzle continued to fascinate me, perhaps because several of the rough-hewn stories contained seemingly autobiographical references to pioneer days in Washington County, Pennsylvania— an area that I was familiar with. During the summer of 1975, I visited Washington, the county seat, and there made the crucial connection: a reminiscence in one of Bancroft's narratives matched almost word for word an anecdote attributed (in Earle R. Forrest's *History of Washington County*) to William Darby, a noted nineteenth-century geographer.

With this information in hand, I consulted the *Dictionary of American Biography* and there found Frank Edward Ross's sketch of Darby, which indeed cites his authorship of the *Casket* tales.

Since the *DAB* entry comprised the only extant scholarship on Darby, I began to explore library and archival resources, with the idea of developing an essay on his magazine narratives. Surprisingly, the research began to yield results far beyond my expectations: I unearthed dozens of Darby's letters and located more than a hundred articles, letters, and sketches that he had contributed to newspapers and journals of the time. I soon realized that his career as a magazinist formed but part of a larger and more engrossing story—his life and labors as a self-educated man of letters. I began to recognize as well that the biography of this frontier traveler seemed to epitomize the era of westward movement itself and provided a personal vantage point on virtually the entire period between the founding of the republic and the Civil War.

Because Darby's own interests were so diverse, I have in writing this book necessarily ventured into fields of learning less familiar to me than American literature. If I have taken what seems to be a narrow, summary view of certain historical or geographical questions, it is because I appreciate their breadth and complexity too much to interrupt the biography with incidental theorizing upon such matters. I have at all times attempted to keep my subject directly in view and to allow his actions and beliefs to illuminate the concerns of the period by individual example.

I am happy to acknowledge here the assistance I have received in preparing this study. A 1976 LSU Summer Research Grant enabled me to gather much of the primary material on Darby in Philadelphia and Washington, and a 1977 NEH Summer Stipend permitted me to complete much of the first draft during a summer sojourn in North Carolina. For these awards, sincere thanks.

I feel a particular gratitude for the kindness of two distinguished professors. The late Arlin Turner, formerly James B. Duke Professor of English at Duke University, saw me through a dissertation on Poe, got me to Baton Rouge, and followed my apprentice years with interest, repeatedly demonstrating his boundless generosity and concern. Lewis Simpson, Boyd Professor of English at LSU, has provided wise counsel, encouragement, and countless recommendations in my behalf. Others on the LSU campus have lent welcome support as well. John Loos, of the department of

History, and Milton B. Newton, of the department of Geography and Anthropology, have both had a long-standing interest in Darby and have furnished different kinds of assistance. The staff at LSU Press has been both eminently professional and humanly considerate at every stage of the publication process. The College of Arts and Sciences at LSU kindly helped to defray typing costs at one juncture. And my colleagues in the English department have sustained their benign perplexity about Darby and thus incited me to finish the book.

Elsewhere, at the Historical Society of Pennsylvania, Tony Roth was extremely helpful; at the Geography and Map Division of the Library of Congress, James Flatness arranged the map reproductions; at the Western Pennsylvania Historical Society, Helen Wilson and William F. Trimble provided valuable assistance. Dr. James T. Herron, Sr., of Canonsburg, Pa., supplied information about Jefferson College, and Gerald W. Palmer of Steubenville, Ohio, played a brief but vital role in the identification of Mark Bancroft.

Finally, my wife Mary Ann deserves credit for more than I could readily acknowledge here. She has understood the compulsions of Darbology.

# THE
# ASTONISHED
# TRAVELER

# Introduction

A few days after the Battle of New Orleans, the commander of the American army appointed a thirty-nine-year-old Louisiana volunteer to the position of topographical engineer. The general had learned that the lean, itinerant surveyor possessed extensive knowledge of the coastal marshland and so assigned him the task of sealing off the bayous south and east of New Orleans against another invasion. That invasion never came, but the engineer performed his duties so capably that the general inscribed a formal commendation—a note destined to figure in a political skirmish between the two many years later. The events at New Orleans marked a turning point in the careers of both men: the surveyor, William Darby, shortly embarked on a new venture as a writer and in the ensuing years won a reputation as an authority on the West; the commander, Andrew Jackson, achieved immediate fame by defeating the British and subsequently became the first son of the frontier to inhabit the White House. And the great victory, which so altered the lives of both men, marked the opening of a new era of westward movement. Waged, ironically, two weeks after a peace treaty had been signed at Ghent, the battle dramatized the indomitability of the pioneer spirit, and it effectively ended British influence in the economic life of the frontier. The action of January 8, 1815, also made plain to the world our determination to settle and defend the West and fostered the belief, soon to become a popular philosophy, that our national destiny was linked inextricably to westward expansion.

The quarter-century following the war's end brought sweeping changes in domestic affairs. The Lake Plains and Gulf Plains, still largely unsettled through fear of Indian attacks or through discouraging rumors of poor soil, saw a massive influx of pioneers. Ray Allen Billington observes: "Those two areas were inundated by a surging tide of westward-moving frontiersmen who swept forward in the greatest population movement the nation had known. By 1850 the peopling of the eastern half of the continent was completed."[1] Precipitated by wild speculation in western land, the Panic of 1819 slowed the migration for several years, but internal improvements, such as the building of government roads, the introduction of river travel by steamboat, and the completion of the Erie Canal, opened new routes to the West and restored the flow of humanity to its earlier level. The configuration of the nation changed as a result—Indiana, Mississippi, Illinois, Alabama, and Missouri all entered the Union in the space of seven years. But no development more clearly reflected the increasing dominance of the West than the election of Andrew Jackson in 1828. The country had, in effect, embraced the frontier life, choosing as president a lean military leader known to despise Indians and political adversaries with equal intensity. His name would soon become synonymous with a populist version of democracy supported across the American frontier.

But the rise of the West evidenced itself in other ways. Everywhere except in New England—which opposed westward expansion to protect its political hegemony—the resources, opportunities, and conditions of frontier life became a virtual preoccupation in the public prints after the war. Booksellers offered an astonishing selection of travel narratives, guides, gazetteers, and chronicles, many of which were designed to encourage settlement in the West. Ralph L. Rusk's admittedly incomplete bibliography lists over two hundred such works, which appeared between 1815 and 1840.[2] Books like John Melish's *Traveller's Directory through the United*

1. Ray Allen Billington, *Westward Expansion: A History of the American Frontier* (New York, 1949), 290.
2. Ralph L. Rusk, *The Literature of the Middle Western Frontier* (New York, 1926), II, 96–144.

*States* (1815) or Samuel R. Brown's *Western Gazetteer* (1817) promised "correct" geographical information, while Henry Schoolcraft's *Narrative Journey of Travels* (1820) and C. S. Rafinesque's *A Life of Travels and Researches* (1836) furnished scholarly insights into Indian life and western flora and fauna. Popular authors tested the market for frontier narratives: Washington Irving penned *A Tour of the Prairies* (1835), Charles Fenno Hoffman contributed *A Winter in the West* (1835), and Robert Montgomery Bird produced *Peter Pilgrim: or a Rambler's Recollections* (1838). Periodicals such as *Niles' Weekly Register* and the *American Journal of Science* published countless essays on the climate, topography, and resources of the West; their readers sought not only accurate information but a sense of the risks and benefits of emigration, reliable fact as well as compelling motive. The massive population shift of the twenties and thirties confirms that they found the latter, at least.

However, travelogues and guidebooks satisfied only part of the popular craving for writing about the West; frontier poems and tales offered a needed imaginative perspective. Strictly speaking, American literature had concerned itself with frontier subjects from the earliest narratives of settlement and Indian captivity, and before the close of the eighteenth century, Charles Brockden Brown had demonstrated the fictive uses of western material (Indian hostilities and wilderness scenes) in *Edgar Huntly* (1799). But James Kirke Paulding's narrative poem *The Backwoodsman* (1818) seems to have heralded the vogue for western fiction and verse which followed the War of 1812; Paulding demonstrated that even a genteel Knickerbocker could find inspiration in pioneer experience.[3] No work had a more galvanizing influence on this literature, however, than James Fenimore Cooper's *The Pioneers* (1823), which sold thirty-five hundred copies on the first day of publication. Cooper's timing could not have been better; to a people moving westward, he offered a genuine western hero, Natty Bumppo, whose further exploits soon followed in *The Last of the Mohicans* (1826) and *The Prairie* (1827). Other writers followed Cooper's example: Timothy Flint, the peripatetic clergyman-author, pub-

3. Prior to *The Backwoodsman*, Paulding had published two frontier narratives in the *Analectic Magazine*: "The Lost Traveler" (August, 1814) and "The Adventures of Harry Bird" (October, 1815).

lished three fustian romances of the West, of which *George Mason, the Young Backwoodsman* (1829) made fullest use of pioneer themes. The more talented James Hall founded the *Illinois Monthly Magazine* in 1830 as a repository for western writing; his own narratives, collected in *Legends of the West* (1832) and *Tales of the Border* (1835), enjoyed wide circulation in eastern periodicals and gift books. Emulating Cooper's methods, William Gilmore Simms attempted to depict the frontier of the Old Southwest in *Guy Rivers* (1834), *Richard Hurdis* (1838), and *Border Beagles* (1840), while Robert Montgomery Bird countered the idealized Natty Bumppo with Nathan Slaughter, a crazed Indian-hater, in *Nick of the Woods* (1837). Elsewhere, William Joseph Snelling compiled his *Tales of the Northwest* (1830), and Caroline Kirkland realistically portrayed the settlement of the Michigan territory in her novel *A New Home—Who'll Follow?* (1839). Collectively the writers of the twenties and thirties produced a body of fiction which reveals much about contemporary attitudes, values, and assumptions concerning the West.

Thus while some authors roamed the western backwoods, publishing factual accounts of what they discovered, others ranged the frontier of the imagination, tracing its image in poetry and fiction. But only a handful of writers—Flint, Hall, and Bird among them—explored the West in both ways. Among this singular group was William Darby (1775–1854), a self-educated man of letters whose life and works—particularly his geographical studies and frontier narratives—form the subject of this study. It would be difficult to find a figure more representative of the postwar passion for knowledge of the West than Darby, who, after his service under Jackson, then launched his career as an author with *A Geographical Description of the State of Louisiana.*[4] In the next thirty-five years, he published more than fifteen volumes of history and geography (including five popular gazetteers), achieving prominence as an authority on the West. Now remembered primarily for his *Emigrant's Guide to the Western and Southwestern States and Territories* (1818), *A Tour from the City of New-York, to Detroit* (1819), and his study of Louisiana, Darby's various labors as an explorer,

---

4. (Philadelphia, 1816). Full publication information for Darby's major works (books, gazetteers, etc.) is given in the Bibliography.

writer, lecturer, and editor reflected his unflagging interest in "the Western country."

Reared in the Pennsylvania backwoods amid incessant Indian warfare, Darby participated in the westward movement and witnessed the astonishing transformation of the West from "a wilderness to a cultivated garden."[5] His various geographical publications enabled a generation of readers to form a more enlightened view of the American frontier. However, not until 1829, when he embarked on a series of border tales for the *Casket* and the *Saturday Evening Post*, did Darby begin to explore the frontier of the imagination. Drawing upon boyhood recollections, oral traditions, and recorded facts, he endeavored to reconstruct for the reading public of the Jacksonian era episodes from what he called "the Heroic Age."[6] These narratives (two of which are reprinted in this volume) complement and illuminate the even more engaging tale of Darby's career as a frontier traveler and man of letters, which until this time has never been told.

5. "Preface," *A Tour from the City of New-York, to Detroit* (New York, 1819), v. The phrase appears in several other works.

6. William Darby to Lyman C. Draper, August 1, 1845, in Pittsburgh and Northwest Virginia Papers, VIII, Draper Manuscript Collection, State Historical Society of Wisconsin, Madison, Wisconsin. All subsequent references to Darby's correspondence with Draper will be to this volume of the Draper Manuscript Collection.

# I. The Life of William Darby

# 1. *From Swatara Creek to "the Western Country"*

William Darby's origins were indeed humble. His mother, Mary Rice, came from Ireland in 1765 and settled with her first husband in Derry, a small farming community east of Harrisburg, Pennsylvania. Some time after the birth of a son in April, 1770, Mary's husband died or deserted her—the exact details remain obscure—leaving her the task of raising a child and providing for herself as well. In such circumstances she met Patrick Darby, a taciturn, hard-working tenant farmer who had come from Ireland in 1772. Drawn together by mutual regard and common poverty, the two were married on November 19, 1774, in a rented cabin on the John Dixon farm. Within a few weeks, the woman discovered that she was pregnant; in a simple cabin beside Swatara Creek she gave birth to a son, named William, on August 14, 1775.[1]

Throughout his life, Darby recalled these early years in the Derry settlement as an idyllic time: he played with other children along Swatara Creek, he was "dandled" on the knee of kindly John Dixon, the family's landlord, and he "learned to read a little" at the Derry schoolhouse, under the tutelage of old John Hutchin-

---

1. The primary sources of biographical information about Darby are his letters to Dr. M. L. Dixon of Winchester, Tennessee. The first, written February 20, 1834, was reprinted in William Henry Egle, *The Dixons of Dixon's Ford* (Harrisburg, 1878), 3–7, and will be hereafter referred to as the *Dixon's Ford* letter. The second, written April 18, 1834, was reprinted in *Notes and Queries Historical and Genealogical*, ed. William Henry Egle (Harrisburg, 1894), I, 33–41, and will be hereafter referred to as the *Notes and Queries* letter.

son. In "Mark Lee's Narrative" (*Saturday Evening Post*, May 24, 1834), he recollected the teasing that Derry children gave to Eneas Grimes, an itinerant weaver supposedly enamored of Jane Montgomery, "a very corpulent Irish woman." Local figures furnished models for William's moral education: the Reverend John Roan, who died six weeks after William's birth, was praised as an example of piety and steadfastness; Robert Dixon, son of the landlord, fell at the Battle of Quebec in December, 1775, and thus became "the first martyr of the Revolution" in the minds of Derry residents. Darby also retained indelible impressions of certain dark scenes, such as the burial of his infant sister Arabella in the Derry churchyard and his last glimpse of grandfatherly John Dixon in 1780. Writing to an actual grandson of Dixon in 1834, Darby remarked: "Well do I remember the very last time he ever walked out of his house and sat down on a bench—in a few days he ceased to be seen of men."[2]

One other scene from the Swatara period would never be forgotten: the wild evening in the winter of 1780/81 when Patrick and Mary Darby resolved to leave the Derry community and remove to the frontier across the Alleghenies. William reconstructed that simple but eventful conversation in "Mark Lee's Narrative":

> Though a child, I might say an infant of five years of age, well do I remember while playing on the floor with one brother older and one younger, on a very stormy night of the winter 1780–81, that after some conversation between my parents respecting their friends already gone there, that my father exclaimed, half in jest, I believe, "Why can't we go to 'the Western country?'" My mother replied in the same playful careless strain, "Why not?"

The offhand proposal soon became a definite plan; in late spring, 1781, Patrick departed for the West to find land and prepare for the relocation of his growing family. He crossed the mountains and reached Washington County, Pennsylvania, where he was welcomed by William Wolf, an "intimate acquaintance" from the Swatara area. On Wolf's farm Patrick planted a crop; he also cleared land along Buffalo Creek and began to construct a cabin.

2. *Dixon's Ford* letter, 3.

After the harvest he returned to Derry to gather up his family and meager possessions for the westward journey. Leaving behind their pastoral life on the Dixon farm, the Darbys set out for the frontier in late October, 1781, accompanied by the John Torrence and Matthew McGreggor families.[3]

To the six-year-old William, the journey to "the Western country" must have seemed a splendid adventure; it evidently engendered a lifelong passion for frontier travel, and it became the vital point of reference in all of his subsequent writing about the West. For the adults, however, difficulties soon developed: less than one hundred miles from the Susquehanna, their wagons had to be abandoned on a tortuous mountain pass. Mary Darby, then nursing an infant daughter, found the two months of muddy roads and makeshift lodging an excruciating ordeal. Beset by fears of Indian massacre, she pleaded with Patrick to return to the security of Derry. But the "slow moving little caravan" continued to move west, encountering horsemen bound either for the backwoods or the East—some bearing frightful reports of Indian raids, others carrying news from the South, where the American forces had surrounded the British at a place called Yorktown. One October day a regiment of soldiers, headed for Virginia, passed along the road, giving William his first view of "the parade of war." But each day the war with Britain seemed more distant; the family moved closer to a theater of violence known to them through captivity narratives and the testimony of those initiated in Indian warfare.

"The season was rainy and unpleasant," Darby later wrote, when the family reached Red Stone (later renamed Brownsville) on the Monongahela River in late November. They rested a few days at Chaffinch's tavern and ferry-house, a rough-hewn log structure of two rooms, through which passed a "constant stream" of sodden travelers on "a pavement of mud." Crowded with pioneers and local "loiterers," the tavern bristled with backwoods chatter about hunting, Indian raids, and the Revolutionary

3. In the *National Intelligencer*, November 6, 1851, Darby described his family's removal to the West, which he there says began in early December, 1781—a date at odds with all of his previous accounts.

War. If "Mark Lee's Narrative" can be trusted as an accurate depiction of the experience, the fraternizing at Chaffinch's was interrupted by the arrival of a rider who announced the dramatic news of the British surrender. But the jubilation was short-lived, as Darby explained in "Reminiscences of the West" (*Casket*, December 1834): "The capture of Cornwallis and his army had electrified the country to the utmost border. It was a story told and retold, but momentary enthusiasm was finally cooled by reflection. Another and far more terrible enemy hovered on the frontier, and . . . scarce a day passed without the report of murders heard."

The proliferation of such reports intensified the dread of Mary Darby; even Patrick felt some anxiety as the family prepared to cross the Monongahela. In melodramatic phrases William later recalled the scene: "On a heavy autumn day, the 25th of November 1781, my mother with a beating heart and face bedewed with tears, three children at her side and one at the breast, crossed the Monongahela; my father, a man never a victim to idle fear, could not entirely conceal apprehensions for his sacred charge."[4] The crossing of the river, which Darby magniloquently called "the Rubicon of that day," marked a symbolic passage into untamed country. The family was met at Gillespie's ferry-house on the opposite shore by William Wolf, who had come to escort them to the Buffalo settlement. Wolf helped to reassure the "terrified" Mrs. Darby by entering into "an animated description of the West, making a mere thing of shadows of Indian war." Mounting the hilly road to the wilderness, the Darbys looked back nostalgically "toward fields they were never to revisit," but they found an unexpected consolation and strength in contemplating the "savageness" of the landscape before them. Impeded by harsh weather, the family made numerous stops west of the Monongahela, finally reaching the area of Catfish Camp (now Washington, Pa.) on Christmas Day, 1781. There they took up temporary residence at William Wolf's farm.

Events of the next three months introduced the Darby family to the extremes of felicity and terror inherent in pioneer life. On January 1, 1782, they celebrated the New Year at a dinner for local

4. "Reminiscences of the West," *Casket* (December, 1834), 541. Further details of the journey to Catfish Camp are derived from this sketch.

families, held in the stockade fort of Jacob Wolf.[5] The occasion marked a reunion of the Darbys and old friends from the Swatara area who had moved west earlier. Shortly after this gathering, at the first break in the winter weather, Patrick resumed work on his cabin, which was completed and occupied in late February—an event typically accompanied by housewarming festivities.[6] But the family remained in the cabin only a few days; in early March, while Patrick was constructing a rail fence, a neighbor named Patty Jolly brought him news of the "savage murder" of a local man named Captain Hawkins. William Darby reconstructed that scene, reflecting the precariousness of domestic life on the frontier, in "Reminiscences of the West":

> This young woman came to my father where he was making fence, whilst myself and little brother were amusing ourselves by his side. I think I see him now, as he stood for at least a minute looking steadily at the young woman after she told in a very earnest manner 'Oh! Mr. Hawkins has been killed.' He had raised a fence-rail, and held it balanced in his hands while the dreadful tidings was communicated and for some time afterwards, but finally dropped the rail, cast a despairing glance over his hard labor, returned to the cabin, and in two hours more we were all on our way to Wolf's Block-House; our cabin was never again seen by one of the family.

The Darbys had reached the frontier—as they quickly discovered—at a period of intense hostility between Indians and pioneers.[7] In 1780, a band of Iroquois braves had launched an invasion of the country west of the Monongahela in retaliation for a militia campaign against their upstate villages the preceding year; in April, 1781, Colonel Daniel Brodhead led a large army from Fort Pitt against the previously friendly Delaware Indians at Coshocton, a foray characterized, according to one historian, by "excessive cruelty in the killing of captured Indians."[8] A series of bloody reprisals by the Indians in late 1781 and early 1782 brought

5. William Darby to Lyman C. Draper, January 1, 1846, in Draper Manuscript Collection.

6. See Joseph Doddridge, *Notes on the Settlement and Indian Wars of the Western Part of Virginia and Pennsylvania* (1824; rpt. Parsons, W. Va.: McClain Printing Co., 1960), 106–108.

7. See Randolph C. Downes, *Council Fires on the Upper Ohio* (Pittsburgh, 1940), 248–76.

8. *Ibid.*, 265.

the Pennsylvania border to such a frenzied pitch that in March, 1782, as the Darby family took shelter "from the fury of savages" in Jacob Wolf's fort, a militia company commanded by Colonel David Williamson set out for the villages of the Moravian Indians along the Tuscarawas River in Ohio. The militiamen believed that marauding Delawares had used the villages of their Christianized brothers to stage their attacks on western Pennsylvania; they were determined to punish the Moravians for their supposed complicity. That the Moravians were a gentle, peace-loving tribe mattered little to the frontiersmen, who sacked and burned the towns of Salem, Schoenbrun, and Gnadenhutten, methodically executing—with tomahawks, knives, spears, and mallets—more than ninety Indians at Gnadenhutten on March 8, 1782. The massacre placed a brand of ignominy on Colonel Williamson, who ironically had hoped to avert bloodshed but failed to control his furious troops. In later years, Williamson became an intimate friend of the Darby family; in his sketch "Reminiscences of the West" Darby attempted to vindicate one of his boyhood heroes.[9]

Believing that the cabin on Buffalo Creek was too remote for their own safety, Patrick moved his family in late March, 1782, to a farm in the Catfish settlement owned by Alexander Reynolds. They were living on the Reynolds farm when Colonel William Crawford led his ill-fated militia expedition against the Wyandot and Delaware Indians along the Sandusky River. Composed of recruits from Washington and Westmoreland counties, the army left Mingo Bottom on the Ohio River in late May, 1782, confident that they would decimate the Sandusky tribes as they had the Moravians. But their movements were watched closely by the Indians, who, with the help of British troops from Detroit, routed and scattered Crawford's men on June 5, 1782; Crawford himself fell into the hands of the Delawares, who tortured and killed him to avenge the atrocity at Gnadenhutten. News of the defeat reached Pennsylvania within days; so too did reports of local men

9. In that sketch, Darby wrote of Williamson, "This man was really anxious to save the Moravians, and had their fate depended on him, not a drop of their blood would have been shed. He was as most militia officers ever are, utterly powerless to stem the violence of men they nominally command." For further details, see my essay, "Glimpses of the 'Heroic Age': William Darby's Letters to Lyman C. Draper," *Western Pennsylvania Historical Magazine*, LXIII (January, 1980), 44–46.

killed in the conflict. Years later William Darby described another scene imprinted on his memory: "James Reynolds came running to my mother exclaiming, 'Jamy Workman is killed.'"[10] But their neighbor James Workman was not dead; he had been separated from his regiment and arrived in Catfish a few days after the announcement of his death. Years later Darby wrote to the frontier historian Lyman C. Draper concerning this border warfare: "Young as I was the incidents were too heart moving and impressive to be forgotten."[11]

One consequence of border warfare was the emergence of a new cultural hero, virtually a new breed, known as the "hunter-warrior." This figure, the most important prototype of which was Daniel Boone, moved between two worlds—the white man's and the Indian's—feeling fully at home, perhaps, in neither. He was, as Henry Nash Smith has pointed out, both an emissary of civilization and a fugitive from it, a self-reliant hero who defended the pioneers by adopting the tactics and often the savagery of the red man.[12] In a letter to Draper, Darby once explained the essential role of the hunter-warrior: "The value of such men as Jonathan Zane, Lewis Wetzel, Henry Jolly, Samuel Brady, and many more, of the heroes of those days of danger and blood, was not made up of what they did, in accidental cases to form and adorn a Tale, but in their watchfulness and in the fact that their names were known and terrible to the savages. It is no risk to assert that for many years of their lives, these brave men were sword and shield to a very extensive frontier."[13] The border narratives penned by Darby in the 1830s, particularly works like "The Hunter's Tale; or, Conrad Mayer and Susan Gray" (*Casket*, November 1831), reflect his enduring admiration of the prowess of the hunter-warrior.

While he was still "very young," probably in 1783 or 1784, Darby actually met one of these backwoods heroes, the redoubtable Samuel Brady. In the absence of her husband, who had departed on a scouting expedition, Mrs. Brady had taken up resi-

10. Earle R. Forrest, *History of Washington County, Pennsylvania* (Chicago, 1926), I, 192.
11. William Darby to Lyman C. Draper, August 1, 1845, in Draper Manuscript Collection.
12. Henry Nash Smith, *Virgin Land: The American West as Symbol and Myth* (New York, 1950), 54–63.
13. Darby to Draper, August 1, 1845, in Draper Manuscript Collection.

dence at the home of her uncle, Andrew Swearingen. Darby and his mother happened to be visiting the Swearingen family when a report of Brady's death reached the community, producing the memorable incident described in a letter to Draper:

> My mother and the Aunt of Mrs. Brady were using all their efforts to calm the distress of the young and devoted wife. Boy as I was my feelings were most powerfully excited, and I now seem to behold and share the scene. When all was in a condition which you are more able to conceive, than I am to describe, the front door was darkened, and by the stalwart form of Brady. The suddenness of the change was really oppressive to all, but almost too much for his young wife.[14]

Darby's unexpected use of the words *darkened* and *oppressive* may be an unconscious reflection of the awesome, almost terrifying presence of the hunter-warrior. Brady's transformation into a ruthless Indian-hater occurred when his brother and father were murdered in 1778 and 1779, respectively; swearing "vengeance against the whole race," he developed extraordinary skills as a scout and Indian fighter, frequently operating in disguise.[15] His effect on the young William Darby may be judged by the tribute paid to him in an early narrative, "The Sioux Chief" (*Saturday Evening Post*, October 10, 1829).

After the bloodshed of 1782, a period of relative calm ensued on the Pennsylvania border. The withdrawal of British support following the peace pact of 1783 forced the Indians to negotiate with the American government, to whom they ceded huge tracts of tribal lands in a series of one-sided treaties. The relaxation of hostilities enabled the Darby family to move into a new residence on "Officer's farm" owned by James Brownlee, in early April, 1783. Here Patrick's agricultural labors met with modest success, for two years later he was able to purchase a farm from Thomas Goudy, at last becoming the owner of the land he cultivated.[16] William had by this time begun to assume a small share of the farmwork, though he relished most the free time in which he

14. *Ibid.*, April 11, 1850.
15. Wills De Hass, *History of the Early Settlement and Indian Wars of Western Virginia* (1851; rpt. Parsons, W. Va., 1960), 382–83.
16. William Darby to Wills De Hass, March 20, 1850, reprinted *ibid.*, 328.

could develop his reading skills. His desire for knowledge grew with his physical growth; Darby wrote, "Every book I could procure, I read, and was aided by a tolerable good memory." In 1785 he encountered the first geographical work he had ever seen, an unidentified text which contained an intriguing, uncompleted map of the lower Missouri River. We can readily imagine that Darby's passion for topography grew from his childhood fascination with the serpentine contour of the river and the mysterious terra incognita toward which it stretched. He also pored over William Sewel's *History of the Quakers*, "the only work on any branch of general history" he had been able to obtain. But the scriptures furnished his primary reading material; Darby claimed that by the age of twelve he had read the Old Testament five times, "and many parts ten times over."[17]

Much of his interest in scripture apparently came from the influence of the Reverend John McMillan, in whose meeting house at Pigeon Creek the boy heard his first sermon in 1782. McMillan, a Presbyterian clergyman, played a leading role in the religious and educational life of western Pennsylvania, founding several congregations, an early Latin school, and an academy which later became Washington and Jefferson College. To the young William Darby, he seems to have represented the saintly counterpart of the warrior-hero—a man of faith, peace, and learning. McMillan appears prominently in Darby's narrative "The Wedding" (*Casket*, July 1836), and a sketch revealing his powerful effect on the boy introduces "Reminiscences of the West": "Thin, spare, pale, and solemn was his visage; his person slender, even light, and his whole figure contrasting and not according with the stern aspect of all around him. His eye however appeared as lamps in the gloom of night. Amid grim war this man seemed a spirit of peace; a spirit of reassurance in an hour of peril he certainly was." The sermon which so stirred Darby was delivered on the "dark and terrible" occasion of the militia's return from the Moravian villages. Interestingly, this "spirit of peace" chose for his text verses from Isaiah (LIV:2,3) which seemed to legitimize the slaughter of Indians and the appropriation of their land: "'Enlarge the place of

---

17. Information about Darby's early reading comes from the *Notes and Queries* letter, 36. He refers to the geographical work in "Mark Lee's Narrative."

thy tent, and let them stretch forth the curtains of thy habitations: spare not, lengthen thy cords, and strengthen thy stakes; For thou shalt break forth on the right hand and on the left; and thy seed shall inherit the Gentiles, and make the desolate cities to be inhabited.'" Darby felt that McMillan took a "rational" view of the border conflict by predicting "the certain ruin of the Indian, and the equally certain spread of the white race." We see in the remark how fully the concept later known as "manifest destiny" depended on the assumption of white supremacy and how oblivious the frontiersman was of the inherent rights of the red man.

Though Indian attacks became less frequent after 1782, sporadic incidents west of Catfish reminded settlers of continuing Indian resentment. In late summer, 1787, the Darby family experienced two horrifying episodes in the space of a few weeks. In August, an old friend named Mr. Crow paid one of his frequent visits to the Darby home, staying most of the day probably to help Patrick with farming chores. A few days later, two of his daughters were captured along Wheeling Creek and brutally murdered by Indians; a third daughter, Christina, escaped and provided an account of the butchery:

> One of the Indians, after a short Parley, began to Tomahawk . . . one of my sisters, Susan by name. Susan dodged her head to one side and the Tomahawk took effect in her neck, cutting the large neck vein, the blood gushing out a yard's length. The Indian [who] had her by the hand jumped back to avoid the blood. The other Indian then began the work of murder, on my sister Mary-Jenar, as she was called. I gave a sudden jerk and got loose from the one who held me and ran with all speed, and took up a steep bank, gained the top safe . . . and hid myself in the bushes near the top of the Hill. Presently I saw an Indian passing along the hill below me. I lay still until he was out of sight. I then made for home.[18]

Shortly after the murder of the Crow girls, tragedy befell another family close to the Darbys, this time involving the sons of a Mr. Becham. William later recalled the arrival of his half-brother, who

18. This account was given to Darby's old friend, Lewis Bonnett, Jr., in 1846, by the woman in question. In early 1847, Bonnett reported the story to Darby, who conveyed it to Draper in a letter of April 10, 1847 (Draper Manuscript Collection).

carried the fateful news: "I had a half-brother, five years older than myself, and while life remains, I must remember his return home and communicating to his parents the murder of two sons, and the scalping and tomahawking of a third, named Thomas, who survived."[19] To the Darbys and their neighbors, the events of 1787 signaled a new epoch of violence and illustrated once more the tenuous nature of frontier existence.

Almost as worrisome as the Indian threat was the poverty of the Darby family; though Patrick somehow managed to buy a farm and property, he faced constant financial woes. With affecting simplicity his son later wrote to a Derry descendant, M. L. Dixon: "My father began and ended by being poor."[20] As a consequence, the Darby children received little formal schooling; William identified an obstacle to his "intellectual advance" in remarking, "I was, from the poverty of my parents, compelled to labor more as my bodily strength increased."[21] The exact cause of Patrick's economic difficulties is unclear, but a likely factor was the growth of his family. Four children crossed the mountains to Washington County in 1781; on the frontier, Mary gave birth to two sons and two daughters. Of this brood only four survived to adulthood—William, his half-brother, his sister Nancy, and his brother Patrick H.—but for about a decade, the elder Patrick struggled to support a family of at least six children. Perhaps due to mounting debts, he sold his farm in early 1793, moving his family to the vicinity of Wheeling. In the autumn of that same year, the eighteen-year-old William received his parents' permission to leave home.

Though his own learning had been "picked up along the Lanes, Highways, and Commons of life," Darby opened a school in Wheeling in late 1793. He explained his qualifications for and attitudes about teaching in an 1834 letter to Dixon: "If I was ignorant, I can say without boast that I had outstript most of my neighbor boys, [and] of course could teach them. Tho' in many respects very irksome business, teaching was of invaluable benefit to me. I had the mornings, evenings, and spare days to myself, and as far

19. De Hass, *History of the Early Settlement*, 328.
20. *Notes and Queries* letter, 36.
21. *Ibid.*

as other means offered, this leisure was used to effect."[22] Darby used much of his free time to satisfy his appetite for knowledge; though Wheeling in 1793 seemed to him the "outer border of civilized life," he managed to procure such works as Rollin's *Ancient History*, Ward's *Mathematics*, and Johnson's *Lives of the English Poets*.

He lived, during part of this period, at the home of Henry Jolly in Wheeling. Jolly had an established reputation as a hunter-warrior, having served in Colonel Morgan's rifle corps during the "Great Campaign" of 1777. His wife, the former Rachel Grice, had in 1775 survived a scalping and tomahawking from an Indian raiding party which massacred the rest of her family.[23] Darby seems to have developed an almost filial regard for Jolly, who in March, 1795, raised a party of pioneers, Darby among them, to establish a settlement in the Ohio country. Writing to Draper in 1845, Darby recalled: "Fifty years on the 20th of last March . . . even then only in my twentieth year, I was one who left Wheeling under the wing of Henry Jolly, to make a settlement where Zanesville now stands. We were driven home by a snow storm."[24] The abortive expedition remained in Darby's memory for good reason: on the morning of March 21, the scouting party for Jolly's contingent encountered a trio of hostile Indians, one of whom was shot by a young Irishman named Denny. His fellow scouts promptly scalped the dead Indian, returning to camp with the trophy. In a note to "Mark Lee's Narrative," Darby wrote, "Though bred from an infant in the frontier, this was the only one of those disgusting memorials of victory I ever actually beheld."

At least one other shocking incident occurred while Darby was a resident of Wheeling. On the evening of September 6, 1794, one of his acquaintances, George Tush, was feeding livestock on his farm when a raiding party fired at him, wounding him in the chest. Perhaps dazed by the shot, Tush stumbled off into the woods, leaving his family unprotected; the Indians quickly captured Mrs. Tush, who (in the words of one historian) "in power-

22. *Ibid.*
23. William Darby to Lyman C. Draper, August 19, 1845, in Draper Manuscript Collection.
24. *Ibid.*

less but quivering agony, [was] compelled to witness the horrid butchery of her innocent children." [25] Later Mrs. Tush, then in advanced pregnancy, was tomahawked to death. Of Mr. Tush, Darby recollected: "The wounded and still bleeding husband and father was brought to our midst [in Wheeling], and placed under the care of George Cookis and his wife." [26] One of Darby's favorite students, Lewis Bonnett, Jr., was then residing in the Cookis home; years later he described to Darby his role in the tragedy: "When Tushes family was murdered, I was nearly a young man, and I was one of the first that went to the House, and saw the whole family laying in the yard tomahawked and scalped, and I helped to bury them." [27] The Tush murders stunned the Wheeling area, for it was popularly supposed that the era of Indian warfare along the Ohio had ended.

During his two-year residence in Wheeling, Darby came in contact with a number of celebrated frontiersmen—"champions of the Heroic Age," as he grandly referred to them in a letter to Draper. [28] In addition to Henry Jolly, who apparently supplied some of the anecdotes related in "Mark Lee's Narrative," Darby counted Lewis Bonnett, Sr. (who appears as a character in the aforementioned tale), Jonathan Zane, and Martin Wetzel (brother of the renowned Lewis Wetzel) as "intimate friends," despite the differences in age. In his later years, Darby came to think of Wheeling as a virtual shrine; when he learned that Draper planned to visit the town in 1846, he wrote: "You will be on Classic ground—on ground made sacred by the tread of heroes." [29] Allowing for Darby's characteristic hyperbole, the remark still suggests a particular reverence for the area, perhaps because of its sanguinary history. Wheeling endured two memorable sieges—1777 and 1782—and Darby came to know many of the participants personally. He was also genuinely moved by the story of the Grave Creek massacre of 1777, in which twenty-one militiamen met their death in an ambush. "Over their grave, for one sepulchre contains their

25. De Hass, *History of the Early Settlement*, 322.
26. *Ibid.*, 329.
27. Transcribed and included in Darby's letter to Lyman C. Draper, April 10, 1847, in Draper Manuscript Collection.
28. *Ibid.*, April 2, 1847.
29. *Ibid.*, August 27, 1846.

bones, I have stood and wept," he wrote in 1835.[30] Darby also associated Wheeling with the end of the Indian wars; on an August day in 1795, he was "one of the party who went over the Ohio river to meet Isaac Zane's family on their arrival after the treaty of Greenville"—the treaty which brought an end to the racial bloodshed in the Ohio Valley. Returning to the area forty years later, Darby declared, "The place and surrounding country, rich in Border History, rich in soft and beautiful scenery—now rich in cultivation, and cheerful to the mind by being the seat of a polished and improving people . . . is of all places, one amongst those which I visit and revisit with most pleasure."[31]

Regardless of his affection for Wheeling, Darby left the area in late 1795 or early 1796, relocating on the Monongahela near Red Stone. In his autobiographical letter to Dixon, Darby noted only that "from Wheeling, in my twenty-first year, I removed to Fayette county, Pennsylvania, and there obtained the perusal of *The Universal History* from Judge Nathaniel Breading. This immense work occupied my every leisure moment whilst I remained in the vicinity of Red Stone, now Brownsville."[32] Darby's serious interest in world history, which led to works like *The Northern Nations of Europe* (1841), may well date from his stay in Fayette County. His source of livelihood is unknown; in view of the fact that we find him five years later serving as a justice of the peace, it seems plausible that he studied law under Judge Breading, perhaps working as his clerk—but this is conjecture. The only event of that period recalled by Darby in later writings was his interview in the fall of 1796 with Thomas Fawcett, an aged tavern keeper who claimed to have shot General Edward Braddock while in his service on July 9, 1755. Fawcett told Darby that he shot Braddock partly to save the army from destruction and partly to avenge Braddock's killing of his brother, Joseph Fawcett, for disobeying orders. Though accepted as truth by Darby, Fawcett's confession (actually a boast) has never been confirmed as fact.[33]

In late autumn, 1796, a journey along the Youghiogheny River

30. See Darby's travel letter from Wheeling, in *Casket* (September, 1835), 493.
31. *Ibid.*
32. *Notes and Queries* letter, 36–37.
33. One account of this meeting appears in "The Great West," *Casket* (September, 1834), 402.

led Darby to the home of an elderly Quaker named Benjamin Gilbert. That meeting, recounted in the narrative "Gilbert and His Family" (*Casket*, April 1835), marked the beginning of an important relationship:

> In the latter part of the year 1796, I was travelling down the Youghiogheny river in Westmoreland county; I was young but suffering under a severe cold, on foot, wet, weary, and unwell. Towards evening I found myself in sight of a plain double log dwelling toward which I proceeded, demanded and received hospitality. By the garb and language I saw that the master of the house was one of the Society of Friends. His features were rough, but the lineaments of his face marked a superior man, which, young as I then was, I soon found to be the case.

About the time of this encounter, Darby moved from Fayette to Westmoreland County; whether he did so on the advice of the Quaker cannot be ascertained, but he soon established himself as a frequent visitor and "intimate" acquaintance of Benjamin and Margaret Gilbert, a childless couple who likely regarded the young man as a son. In Benjamin Gilbert, Darby discovered both a second father and an intellectual guide, whose library—by frontier standards at least—must have been impressive. In his autobiographical letter to Dixon, Darby wrote: "With Mr. Gilbert's books an entirely new species of reading was opened to my mind. From this man I procured the reading of *Montesquieu's Spirit of Laws, Locke's Essay on the Human Understanding, Reed on the Mind, Blair's Lectures on Elocution, Elements on Criticism, by Henry Home Lord Kaimes*, and perhaps the deepest metaphysical work ever written, *Edwards on Free Will*." [34] Under the influence of Gilbert, Darby pursued a course of reading in philosophy and theology which led to serious consideration of a "clerical profession."

Through his mentor Darby also became acquainted with the tenets of the Quaker religion and with the personal history of the Gilbert family, whose captivity by the Indians in 1780 formed the subject of a pamphlet by Benjamin Gilbert—the only member of his family not involved in the ordeal. Darby read the pamphlet with great interest, and upon the arrival of Abner Gilbert, Ben-

---

34. *Notes and Queries* letter, 37.

jamin's half-brother (who took up residence there in 1797), he acquired a firsthand account of the captivity and Indian customs. "Many an hour have I most attentively listened to Abner's plaintold tales," he wrote later, adding that the storyteller "always spoke affectionately of the Indians, and seemed to retain a strong poetic remembrance of the wild, uncertain, wandering and changeful life of his savage captors, and in their manner, afterwards his friends."[35] The gentle Quakers must have seemed strange folk to Darby after his acquaintance with hunter-warriors who had taken scalps. Drawing upon his recollection of Abner's "tales" and an available copy of Benjamin's pamphlet, Darby published "Gilbert and His Family" almost forty years later as a tribute to the courage of the Gilberts and—indirectly—as a monument to the friendship, care, and enlightenment he received from members of the Quaker family.

During 1797 and 1798 Darby evidently continued to search for a direction in his intellectual growth; how he managed to support himself is not known. Nor can we be sure how long he resided in Westmoreland County; the only clue appears in his vague reference pertaining to Benjamin Gilbert: "I remained in his neighborhood some years."[36] In *The Emigrant's Guide*, he speaks of visiting Steubenville, Ohio, during the first week of January, 1799,[37] but whether he returned to the Youghiogheny or resettled on the Ohio River, we do not know. One fact is certain: in June, 1799, his father died in Washington County, and that loss effected a major change in his life.

35. "Gilbert and His Family," *Casket* (April, 1835), 184, 188.
36. *Ibid.*, 183.
37. *The Emigrant's Guide to the Western and Southwestern States and Territories* (New York, 1818), 228.

# 2. Surveyor in the Old Southwest

During his years in Wheeling, William Darby must have spent countless hours rambling along the Ohio River, enjoying the vistas from its mountainous banks and imagining its course through the wilderness toward the mysterious, alluring Southwest. The passing flatboats, crowded with emigrants bound for Kentucky, Alabama, Mississippi, and Louisiana (then still held by the Spanish), awakened an inevitable longing, which the tedium of his teaching duties did little to relieve. And so shortly after the burial of his father in the churchyard at the Chartiers Creek meeting house, Darby resolved to leave the Pennsylvania border country—which had already begun to lose its frontier aspect—to follow the river's current toward a region still largely unsettled. Darby later explained that the death of his father "induced" him to travel, as if that event simultaneously released him from an obligation and dramatized the need to seek his fortune elsewhere. Patrick Darby had removed to the frontier in 1781 believing that it promised an escape from poverty or at least the opportunity for competence. But he died a poor man, and predictably his death aroused the same hope in his son: to find that bountiful frontier where he might prosper as a free man. William Darby sensed that these expansive dreams could no longer be realized in Washington County or the Ohio Valley; already many of the heroes of the border wars had moved on to the Southwest; families which participated in the settling of Catfish Camp had succumbed again to the desire for change. Like the characters described at the close of "The Wedding" (*Casket*, July 1836), Darby himself was in July, 1799, "drawn away by that infatuation which places paradise on the outer verge of civilization."

He was thus doomed by naïve expectation to experience a disillusionment familiar to many who pursued the dream of the West. In midsummer, he boarded a riverboat at Pittsburgh and for the next nine weeks endured a decidedly unromantic passage to paradise: unbearable heat, ravenous mosquitoes, and rampant dysentery made the voyage memorable for just the wrong reasons. His arrival in the Southwest was similarly inauspicious; in *A Geographical Description of the State of Louisiana*, he recalled: "In the months of July, August, and September of 1799, I descended the Ohio and Mississippi to Natchez. That year the yellow fever prevailed in New-Orleans; to the severity of which, fell victims, Manuel Gayoso de Lemos, the then governor of Louisiana, and many other persons, Creoles, and strangers. I arrived at Natchez on the 13th of September, in that country the most deleterious month."[1] We cannot be certain whether Darby had intended to stop at Natchez or simply disembarked there after learning of the epidemic in New Orleans; in any event, he soon settled in a cabin on Pine Ridge, north of Natchez, and lived there for the next eighteen months.

Reared among Irish and Scotch-Irish pioneers in Pennsylvania, Darby must have been fascinated by the cultural diversity of Natchez, whose citizens variously expressed loyalty to Spain, France, Britain, and the United States. In "Caroline Marlow" (*Saturday Evening Post*, March 13, 1830), he endeavored to depict the social life of the town under Governor Gayoso, inserting into the tale portraits of local leaders he had known. He arrived in Natchez during a period of political turmoil; the Congress had created the Territory of Mississippi in April, 1798, and the ap-

---

1. *A Geographical Description of the State of Louisiana* (2nd ed.; New York, 1817), 279–80. All subsequent references to the *Description of Louisiana* will be to the revised second edition.

pearance of President Adams' hand-picked governor, Winthrop Sargent, created an immediate backlash among British partisans, who resisted the authority of the American government.[2] Darby witnessed a subsequent clash between Governor Sargent and Colonel Anthony Hutchins, an outspoken Tory planter. When Hutchins presented himself to take the oath of office as a member of the territorial legislature in 1800, the embattled Sargent stubbornly refused to administer it, charging that Hutchins was a British subject. "The Col. sternly replied that he was," Darby later recalled, "but claimed what could not be denied or refused, citizenship under the Treaty which secured the Territory to the United States."[3] In the political wars of early Natchez, Hutchins actually outlasted Sargent, who was replaced as governor by William C. C. Claiborne after the federal elections of 1800.

At the beginning of the nineteenth century, Natchez teemed with colorful, ambitious men, and Darby numbered several among his personal acquaintances: Hutchins, whom he considered an "honest and honorable man" of "great energy"; Ferdinand L. Claiborne, the younger brother of William and later speaker of the Legislature; and William Dunbar, a distinguished scientist as well as a cotton planter, "long the very first man forming the society of the Natchez region." Darby also discovered living in the area numerous veterans of the border wars of Pennsylvania, Ohio, and western Virginia. For example, he was "no little surprised to find" residing in the vicinity "the very Old Hero Adam Bingaman," whose exploits Darby had heard about in Catfish.[4] One of George Rogers Clark's officers, Captain John Girault, became influential in Natchez politics and won a seat in the territorial legislature in 1802. Curiously enough, even the fabled Lewis Wetzel surfaced in Natchez after serving a prison term in New Orleans for his unwitting involvement with a counterfeiter; the hunter-warrior supposedly died in Mississippi in 1808. None of the former Indian fighters around Natchez aroused more excitement, however, than

2. Robert V. Haynes, "The Formation of the Territory," in *A History of Mississippi*, ed. Richard Aubrey McLemore (Hattiesburg, 1973), I, 174–216.

3. William Darby to Lyman C. Draper, May 21, 1847, in Draper Manuscript Collection.

4. *Ibid.*

Samuel Mason. "Well would it have been for Capt. Samuel Mason if he had fallen with his gallant companions on the field at Wheeling," Darby opined in "Mark Lee's Narrative." A participant in the siege of 1777, Mason drifted into the Mississippi Territory at the turn of the century and for several years led a notorious gang which terrorized the entire area, preying mainly on travelers along the Natchez Trace.

Amid such men, famous and infamous, Darby struggled to make his way in the Old Southwest. How he initially supported himself remains unclear, but his fortunes took an abrupt and perhaps calculated rise when on September 8, 1801, he married a middle-aged widow named Elizabeth Boardman who owned a nearby plantation. He explained the circumstances of their union later to Dr. M. L. Dixon: "I went to Natchez, where, very contrary to my expectations, I married . . . a widow with a family of children and quite handsome property. What led me into this connexion was a similarity of tastes. Like myself, Mrs. Boardman had been her own teacher, and had acquired a fine stock of information."[5] Whether Darby was more attracted to the widow's "fine stock of information" than her "quite handsome property" is a nice question, for which Darby's ingenuous testimony provides the only clue. Elizabeth was in fact the widow of Charles Boardman, a wealthy planter and landowner, and her remarriage to a man at least ten years younger evidently scandalized certain members of the Boardman family.

Through this advantageous alliance, Darby thus found himself transformed almost overnight into a cotton planter, a man of means, and even (in his own mind, at least) a member of the landed gentry. He began to play a modest role in local affairs: by 1802, he had become the captain of a militia company, charged with keeping order in one of the ten districts of Adams County; the same year Governor Claiborne appointed him to serve as one of seven justices or "Esquires" on the bench of the county court. Records indicate that he retained both positions for at least two years. In June, 1803, Darby became the tax assessor for his militia

5. *Notes and Queries* letter, 37. The only extant record of the marriage date is Darby's letter to Draper, April 10, 1850, in Draper Manuscript Collection.

district, and the following year he also prepared the tax list for the district of Ferdinand L. Claiborne.[6] But his involvement in the legal and political life of Natchez, made possible by his marriage to Mrs. Boardman, was also curtailed by a backlash of family resentment; of his conjugal life with Elizabeth, Darby later recalled: "As a wife she was everything I or any man could wish for, but her family involved us in litigation."[7]

His troubles began in June, 1803, when members of the Boardman family took legal action to replace Darby as guardian of Charles Boardman's six children. The move had probably little to do with his fitness as guardian; rather, it represented an effort to place beyond his reach certain funds designated for the raising of the children and several tracts of land held in trust for them. In October, 1803, the family forced an audit of the plantation accounts under Darby's administration, hoping to find evidence of managerial incompetence, and the following year they initiated a reexamination of Boardman's will to challenge the dowry "in lands and tenements" provided in the event of Elizabeth's remarriage. As a result of this litigation, Boardman's children became the wards of John Henderson and William Foster, Natchez attorneys who represented the family's interests. Thus, although the children continued to live with their mother and stepfather, the funds which had once been available to support their nurture were cut off. Despite his status as a justice of the Adams County Court, Darby found himself outmaneuvered by the family lawyers; the entries in the Orphan's Court Records, which testify to the ongoing legal harassment, suggest that Darby came to be regarded as an interloper, unworthy to manage a great estate.[8] What the family thought of Elizabeth Boardman Darby we can only imagine.

In the midst of these difficulties, which drained the couple's resources and patience, a catastrophic fire destroyed a "large and well-filled cotton gin" on the plantation in the spring of 1804. To recoup his losses, Darby was forced to sell seven slaves (three

6. *Transcription of County Archives of Mississippi, No. 2, Adams County* (Jackson, Miss., 1942), II, *passim*.
7. *Notes and Queries* letter, 37.
8. Frequent references to this controversy appear in the *Orphan's Court Records,* 1800–1810, Adams County Courthouse, Natchez, Mississippi.

men, a woman, and her children), some livestock, and much of the household furniture on May 9 for a mere $961. The second blow came a few weeks later: a storehouse full of cotton burned to the ground, necessitating the sale of a valuable slave for $500 on July 1 and then the mortgaging of a "messuage tenement" and another tract of 180 acres on August 27 for $783. These entries from the Conveyance Records in Natchez, impersonal notations in faded ink on a yellowing page, hardly begin to describe the private crisis to which they bear witness. Many years later, Darby tersely remarked to M. L. Dixon: "This double loss involved me in debt, to which I was compelled to yield."[9] Ironically, his first taste of the prosperity and ease which had eluded his father led to colossal debts which would haunt Darby for years.

With the advantage of hindsight, he later wrote, "Had I not lost this property, and been thrown once more on my own resources, I would no doubt have vegetated a Mississippi cotton planter."[10] But financial pressures forced Darby and his wife to give up the Boardman estate, apparently in late 1804. During that bleak period, their principal income derived from rents collected on property still owned by Elizabeth (or "Eliza," as her young husband called her). Darby obtained some additional cash from surveying work, which enabled him to draw upon his mathematical acumen and to refine the mensural skills acquired as the "highway overseer" of his militia district. Planters and land speculators, eager to stake new claims in unsettled areas of Mississippi and Louisiana, hired him to measure and mark tracts of wilderness land. According to a reminiscence in the *National Intelligencer* (December 5, 1848), Darby "first entered on the plains of Louisiana" in October, 1804, on what was probably a surveying mission along Bayou Teche in the Attakapas country. The fledgling topographer would long recall that "most splendid autumn evening" and the rapture inspired by "those immense grassy plains, which, on first and last view, give the soul ideas of immensity only rivalled by an oceanic prospect." Thus it happened that financial disaster led him into the territory which would become the object of his most important

---

9. *Notes and Queries* letter, 37. Information about Darby's property transactions comes from *Conveyance Records*, 1800–1810, Adams County Courthouse, Natchez, Mississippi.
    10. *Ibid.*

geographical study. But in 1804, Darby had no inkling of the herculean labors which lay ahead; he was then simply concerned with gathering information about the land claims he had been paid to establish.

The excursion into Louisiana's fertile prairie evidently convinced Darby that, once again, westward movement might resolve a personal dilemma. Consequently, on July 9, 1805, he and Eliza sailed down the Mississippi to the Atchafalaya River, their boat loaded with all the furniture that could be saved from the clutches of creditors. Their destination was Opelousas, an old French outpost southwest of Natchez, still possessing a frontier ambience and populated mainly by French and Spanish settlers. Accompanied by Eliza's daughter, Maryanne Boardman, the couple found temporary lodging there with John Thompson, who had been appointed as the clerk of the government Land Office in Opelousas; through Thompson's kindness, the bankrupt cotton planter found private surveying work in the area. Shortly after his arrival, Darby also began to study French, a pursuit which no doubt gratified his scholarly nature even as it prepared him for commercial dealings in Opelousas. In October, 1805, he attended the official opening of the Land Office, watching with amusement as government commissioners, barely conversant with new laws and policies, struggled to evaluate the claims of French-speaking citizens whose deeds reflected a perplexing system of entitlement. On November 13, Darby himself acquired a tract of land on the Bayou del Puente for the price of one dollar, having made over an indenture bond of $1,200 to a Frenchman named Michel Bordelon. He hoped to pay for the property with his earnings as a surveyor; over the winter months, he marked off claims in Concordia Parish and the Attakapas country for several land speculators, and he also found work as an assistant to Deputy Surveyor James Gordon of the federal Land Office. By March, 1806, his fortunes had improved so rapidly that Darby ventured into three separate land purchases with a partner, Benjamin P. Porter; the following month, he combined assets with Robert Rogers to secure a tract on the Nez Piqué River. Government records suggest a pattern in Darby's early speculation: he invested small amounts to buy tracts whose titles had been contested, and in nearly every instance his claim of ownership was ultimately rejected in Washington by the federal Land Office.[11]

But in late spring, 1806, his dealings with that agency took on a different aspect. Though his claims had been refused, the young surveyor had made a favorable impression on Clerk John Thompson, and when a deputy surveyor named Christopher Bolling was found guilty of "disgraceful, dishonest conduct," Thompson wrote to Isaac Briggs, surveyor of public lands south of the state of Tennessee, recommending Darby for an appointment:

> In the course of last Fall, I mentioned Mr. Darby to you, as a person, whom I supposed from my acquaintance with him, to be properly qualified in surveying the lands of this District—Mr. Darby has since acted as an Assistant to Mr. Gordon, and has resided a great part of his time in the same house with me:—and from opportunities I have had of observing his conduct, do not hesitate to recommend him to your Notice and attention—His particular wishes, he will communicate to you himself; and I think you may repose in him all necessary confidence.[12]

The young surveyor's letter of application presumably followed, for on May 14, 1806, Briggs wrote from his headquarters in Washington, Mississippi, to revoke the commission of Bolling, appointing Darby in his place. Through this process, Darby became one of six deputy surveyors for the Western District of the Territory of Orleans.

The position brought with it the security of steady work and relatively substantial income; moreover, it provided Darby with information about good land available in the Opelousas area and a better understanding of the procedures for authenticating claims. Two other developments increased the importance of the government appointment: several months after their arrival in Louisiana, Eliza (then in her early forties) presented Darby with a daughter, whom the couple named Frances. The surveyor had barely adjusted to this new responsibility when he learned that his creditors in Natchez were taking legal action to guarantee that Darby would honor the bonds of indebtedness he had signed before leav-

11. See *American State papers, Public Lands* (Washington, 1834), III, 105, 112, 113.

12. *Deputy Surveyors' Letters*, Southwest District, 1805–1810, State Land Office, Records Division, Baton Rouge, Louisiana.

ing Mississippi. In early February, 1807, he journeyed to Natch-ez—making notes along the way on the creeks and bayous he crossed—to face his creditors and plead for an extension of the deadline for payment. As a consequence of the negotiations, six creditors filed a joint indemnity claim on February 11, placing a lien on all of Darby's possessions at that moment—three tracts of land on the Grand Prairie of Opelousas and his personal property: "one small black horse; eleven tables; two which mahogany; one walnut; one maple and the residue cherry; one writing desk; three beds and their furniture; one dozen silver spoons."[13] The creditors each listed the amounts owed to them by Darby, with the total indebtedness exceeding two thousand dollars. Under the terms of this "Letter of Licence," the surveyor had two years to pay off his debts or forfeit his property. Fearing that he would never escape from this financial nightmare, Darby tried to cover himself against complete ruin by utilizing a technical loophole; since the lien ap-plied to all that he then possessed (or might subsequently pur-chase), he shrewdly named his daughter Frances as the owner of those properties which he acquired after early 1807. But the "Let-ter of Licence" effectively destroyed any hopes Darby might have entertained of recovering prosperity through land speculation; to meet his Natchez debts, he was forced to sell two large tracts on the Grand Prairie, retaining only the third, where he lived with his family and two slaves, raising fruits and vegetables to meet their domestic needs.

These financial troubles and land transactions form the personal background of Darby's important surveying activities in Loui-siana. The actual point at which he conceived the idea of pro-ducing a "Map and Statistical Account" of the state cannot be determined with much precision; he gave several dates for the commencement of his project, which apparently developed at first as an unsystematic accumulation of notes and records and then grew into a deliberate, organized study in late 1808 or early 1809. In a sense, his exploration of Louisiana and the Gulf Coast began the day he left Natchez, a date given emphasis in his *Description of Louisiana*, which includes this summary of the scope and duration of his labors: "Between the 9th of July, 1805, and the 7th of May,

13. *Deed Book B*, Adams County Courthouse, Natchez, Mississippi, 292–93.

1815, incredible as it may appear to many persons, I actually travelled upwards of twenty thousand miles, mostly on foot. The range of my peregrinations embraced all the country between the Mobile Bay and the Sabine River; and from the Gulf of Mexico to the thirty-third degree of north latitude."[14] While the extent of his travel does seem "incredible," the numerous and meticulous mileage tables in *The Emigrant's Guide* testify to his accuracy about such matters; it should be added, however, that some of the travel included in this estimate had objectives unrelated to topography. And though Darby emigrated to Louisiana in July, 1805, and carried out private surveying work at various sites prior to his government appointment, he seems not to have begun his actual field investigations as deputy surveyor—that is, his first detailed study of a specific area—until the end of 1806.

Under the scheme devised for the government survey of Louisiana, the Territory of Orleans was divided into quadrants at 31° north latitude and (approximately) 92° 27′ west longitude; the Western District originally included all of the land west of the Atchafalaya River. Gideon Fitz, the principal surveyor in charge of the Opelousas Land Office, designated the ranges (or townships) east and northeast of Opelousas as Darby's responsibility. The geographer later reflected on his assignment in a letter to the *National Intelligencer* (February 26, 1836):

> In the distribution of districts, the tangled woods northeastward of the village and church of St. Landry fell to my share. The region, after leaving the village about four miles, and crossing the Courtableau River, was then, and remains now, one of the most dense, and with an undergrowth of cane and palmetto, interlaced with saw-briar vines, one of the most impassable forests of this continent. As a region for surveying it was worse than unprofitable; but into it I plunged, and obtained what could only be obtained by actual view—a knowledge of the true features of the individual basin of the Atchafalaya which most essentially differ from those of the Mississippi.

Though Darby here complains that his government work was "unprofitable," in fact records show that he was paid at the rate of

14. *Description of Louisiana*, 280.

four dollars per linear mile of surveying, or between $150 and $200 per range. From early 1807, when he filed his first set of field notes, until the end of 1808, he completed the measurement of seventeen ranges. When we allow for additional fees collected for private surveying, it seems probable that Darby's income must have approached $2,000 annually during this period. Yet the work was unquestionably arduous; with his crew of chain carriers (which in 1808 included a man named Mark Lee, whose name would later be given to the hero of "Mark Lee's Narrative"), Darby spent most of 1807–1808 slogging through the swamps of the Atchafalaya basin, exposed to heat, humidity, alligators, and mosquitoes as he hacked through the rank undergrowth to mark township boundaries.

As a consequence of his assignment, the Atchafalaya River became one of the first subjects of independent investigation by Darby. In 1808, he carried out an extensive study of the "raft" or logjam which had obstructed the river since 1778. He calculated that the raft stretched twenty miles from north to south, with at least ten miles of the river blocked from shore to shore by the accumulated timber. Yet even in the midst of such research, the surveyor was responsive to the natural beauty of the scene:

> In the fall season, when the waters are low, the surface of the raft is covered by the most beautiful flora, whose varied dyes, and the hum of the honey bee, seen in thousands, compensate the traveller for the deep silence and lonely appearance of nature at this remote spot. The smooth surface of the river, unoccupied by the raft, many species of papilionaceous flowers, and the recent growth of willow and cotton trees, relieve the sameness of the picture; even the alligators, otherwise the most loathsome and disgusting of animated beings, serve to increase the impressive solemnity of the scene.[15]

Darby explored the raft as a natural extension of his government work, since the obstruction lay in that portion of the river contiguous to the ranges under his purview. But during the summer of 1809, he became involved in a private project which enabled him to expand his knowledge of the river: with the assistance of

15. *Ibid.*, 133.

a guide named Joseph Chatehault, Darby opened the first path between Opelousas and the mouth of the Atchafalaya. Though within one hundred miles of New Orleans, the area traversed by the geographer was then still "a land unknown"; he completed his task "in defiance of all obstacles" and simultaneously collected firsthand information about the lower reaches of the Atchafalaya. Subsequent expeditions took him north of Bayou Rouge to the river's origin at the confluence of the Red and Old Rivers. In a newspaper summary of his Atchafalaya investigations (*National Intelligencer*, February 25, 1836), Darby later concluded: "As early as 1808, I had conceived the design of a Map and Statistical Account of Louisiana, a circumstance which inspired and kept active an attention to the phenomena presented far more lively than could have done the exercise of the ordinary duties of a Deputy Surveyor." Thus it seems evident that his independent work in the Atchafalaya basin directly inspired the more comprehensive scheme of a geographical study of Louisiana.

His close observation of the river and surrounding marshland continued through 1810, partly as a result of government assignments. The area around Opelousas fascinated Darby, and though his surveyor's field notes indicate the mass of tedious details which commanded his attention on the job, his geographical writings reflect a larger vision of the land. Ever regardful of natural boundaries—whether of vegetation, climate, or terrain—he was especially struck by the contrast between the wetlands of the Atchafalaya and the rolling prairie which rises near Opelousas:

> A more rapid and astonishing transition is not conceivable, than between the deep, dark, and silent gloom of the inundated lands of the Atchafalaya, and the open, light, and cheerful expansions of the wide spread prairies of Opelousas and Attacapas. This pleasing and really delightful change is amongst the certain items of reward, that every individual will receive, who passes at any season of the year from New Orleans to either Opelousas or Attacapas. After being many days confined in the rivers, exposed to heat, musquetoes, and many severe privations, to pass in a few minutes from this scene of silence and suffering, to an ocean of light, to view the expanses where the eye finds no limit but the distant horizon, is a delight of

which no anticipation can give an adequate idea. To be enjoyed, it must be felt.[16]

This passage reflects a quality characteristic of Darby's early works: the willingness to express, among scientific observations, the romantic delight, even the astonishment, inspired by the natural landscape. As he wandered through the Louisiana bayou country, the surveyor often imagined himself as a traveler through time as well as space. In another description of the deep solitude of the Atchafalaya swamp, he mused on the fanciful effect of trees "rendered venerable by the long train of waving moss" and remarked: "The imagination fleets back toward the birth of nature, when a new creation started from the deep, with all the freshness of mundane youth."[17] Looking out on the Opelousas prairie, with its thousands of grazing horses and cows, Darby saw "a sea of plenty" and remarked, "If the active horsemen that guard them would keep their distance, fancy would transport us backwards into the pastoral ages."[18] Such passages reveal both a keen eye for detail and a lively, historical imagination which enabled Darby to perceive pictorial allegory wherever he looked.

From all available evidence, the surveyor's travels prior to 1811 were confined mainly to the area within a seventy-five-mile radius of Opelousas, though he probably made occasional journeys to New Orleans by pirogue and riverboat. But in August, 1811, Darby resigned his position as deputy surveyor so that he could complete a full-scale investigation of Louisiana and the Gulf Coast. It may be that this strategy was prompted by recent events: American rebels had captured Baton Rouge from the Spanish in 1810, the United States claimed West Florida, and expansionists in Washington pressed for statehood for Louisiana. These developments shifted national attention to the Territory of Orleans, increased the prospects for a dramatic influx of settlers, and thus enhanced the potential value of a geographical guide to the area. After resigning his government post, Darby apparently devoted several weeks to his manuscript, transcribing notes and consolidating his observa-

16. *The Emigrant's Guide,* 72.
17. *Description of Louisiana,* 136.
18. *Ibid.,* 106.

tions on the Atchafalaya basin and the prairies of Opelousas and Attakapas.

According to a deposition prepared for the U.S. Congress, the geographer then made "an extensive tour of what is now the State of Louisiana" during the "latter part of 1811." [19] On this expedition he followed the Red River to the northwestern corner of the territory and "traversed repeatedly the hitherto most imperfectly known parts" of the border region. He ranged to the northeast as far as Bayou Macon, near the Mississippi, and investigated with particular avidity the banks of the Ouachita River and the vast lands granted to the Baron de Bastrop by the Spanish government in 1795. Upon his return to Opelousas, however, Darby was forced to confront the problem of financing the rest of his planned excursions; according to his congressional deposition, he appealed without success to the Louisiana legislature for monetary support, and on January 14, 1812, he tried to enlist the help of Ferdinand L. Claiborne, then a brigadier general in the Mississippi militia: "Enclosed you will receive a copy of the proposals for my intended work on Louisiana. The success or failure of this attempt will determine whether I will have it in my power to do justice to my Creditors at the Natchez or otherwise." [20] Darby emphasized his desire "to do justice" to his creditors because Claiborne himself happened to be among that disgruntled contingent; perhaps anticipating Claiborne's refusal, the surveyor also expressed his bitterness about the Mississippi town, calling it "a place where I have been the sport and object of so much calumny." Darby's letter thus indirectly reveals the sad fact that three years after the expiration date of the "Letter of Licence," he had not yet fully settled the debts incurred in 1804.

Faced with such discouraging circumstances, the geographer struggled to raise funds for a new expedition through private surveying during February and March, 1812. Then, in early summer, Darby set out for New Orleans and boarded a schooner at Fort St. John; the vessel sailed through Lake Pontchartrain, past the Rigo-

19. U.S. Cong., Senate, *Reports of Committees*, 30th Cong., 1st sess. (Washington: Wendell and Van Benthuysen, 1846), I, Rept. 236, p. 5.
20. William Darby to Ferdinand L. Claiborne, January 14, 1812, in Dreer Collection, Manuscript Division, Historical Society of Pennsylvania, Philadelphia, Pa.

lets, and into the Mississippi Sound. Seizing every opportunity to gather new facts and figures, he measured the depth of the coastal waters at regular intervals, and he pressed the ship's captain for information about the chain of desolate islands protecting the shoreline. When the schooner entered Mobile Bay, Darby probably disembarked to inspect Mobile and then proceeded to his ultimate destination, Fort Stoddert, a post on the west bank of the Alabama River. Why the surveyor journeyed so far to visit what he later termed "a place of little note" remains a mystery; conceivably, he hoped to solicit funds from an officer stationed there, but no evidence of such a plan remains. Darby returned overland to New Orleans in July, suffering greatly from mosquito swarms between Fort Stoddert and the Pascagoula River; he followed the Gulf coastline across the Pearl River to Madisonville, Louisiana, where he boarded a ship for passage across Lake Pontchartrain.

After several days in New Orleans, probably devoted to the search for a patron or some official sponsorship from the newly created state of Louisiana, Darby made his way back to Opelousas. But he remained at home only a few weeks; after transcribing the notes of his journey to West Florida and enjoying an interval of domestic contentment with his wife and young daughter, the geographer set out again in early October, 1812, on what would become his most important single expedition: a survey of the Sabine River. He followed the Red River northwest as far as Natchitoches, an old Spanish outpost, where he enlisted support from two prominent citizens, Edward Livingston and Dr. John Sibley, the latter supplying some of the surveying instruments used on this project. From Natchitoches, Darby and three assistants headed west-northwest toward the Sabine, measuring several different routes to the river and extending the investigation as far north as the "then vaguely known settlement of Bayou Pierre."[21] In November, he returned to the point where, as he calculated, the thirty-second parallel crossed the Sabine and from that site embarked on his momentous journey to the Gulf of Mexico. The river was then flooded and the desolate appearance of the area convinced Darby that he was (as he said in the *National Intel-*

21. *Emigrant's Guide*, 85.

*ligencer*, May 13, 1836) "the first individual who ever reached the spot with any civilized object in view." Indisputably, he was the first professional surveyor and geographer to undertake a study of the Sabine; he described his mode of travel in the same newspaper essay:

> On the slope of the pine hill bordering the lake [which once lay near present-day Logansport, La.], I had a large pine tree felled, and out of its trunk a pirogue constructed. In this pirogue I launched into the Sabine, with three men and our baggage and provisions. From the size of the river there . . . I had no hesitation in supposing its sources at least one hundred miles more remote. My hunter, a man of the name of Wallace, insisted that it was a still greater distance to the headwaters of the river. The channel was at least forty yards in width, and I found no impediment to the navigation, not even fallen timber, and descended the stream to the lowest Indian village, where I exchanged my pirogue for one still larger, and with which I navigated out of the mouth and up the Calcasieu to its upper lake, from whence I returned to Opelousas by land.

Initially, Darby had entertained the idea of tracing the Sabine to its source, but this scheme involved crossing into Spanish territory, and his friends in Natchitoches persuaded him that he would risk "personal danger" if discovered by Spanish troops patrolling the border. While thus restricted by political considerations, the geographer nevertheless produced the first accurate map of the river as it forms the western boundary of Louisiana. The journey from the thirty-second parallel to the mouth of the Sabine took more than a month, during which time the party subsisted mainly on fish and game.

Darby arrived at the head of Lake Sabine on December 20, 1812, and with his men he endured "severe frost" on the following two nights. The dreariness of the season, coupled with the bleakness of the marshland, inspired this lasting recollection:

> No prospect can be more awfully solitary, than that from the mouth of the Sabine. A few trunks of trees thrown on shore by the surf of the sea, and scattered clumps of myrtle, are the only objects that arrest the eye, from the boundless expanse of the gulph, and the equally unlimited waste of the prairie. . . . The deep solemn

break of the surge, the scream of the sea-fowl, the wind sighing mournfully through the myrtle, and a lone deer bounding along the shore, are the only objects that vary the monotony of the scene; the only sounds that interrupt the awful silence of this remote region.[22]

Darby set out for Opelousas on January 3, 1813, following the coastline as far as Calcasieu Pass and paddling to the head of Calcasieu Lake before abandoning his pirogue for the overland trail. Though he had been away from his family nearly three months, the geographer tarried only a few days in Opelousas; urgent business called him on to New Orleans, where he arrived on January 15. Armed with fresh, authoritative information about the Sabine and the western part of the state, Darby doubtless tried to capitalize on this material to gain backing for his "Map and Statistical Account of Louisiana," but again he met with failure. Yet the rapid journey from the remoteness of the western marshes to the bustle of the cosmopolitan city had its reward in historical insight; this passage seemed to Darby a veritable allegory of human evolution:

> A journey from New Orleans to the mouth of the Sabine, exhibits man in every stage of his progress, from the palace to the hut, and inversely. To an observing eye, the rapid transition from the superb mansions of the wealthy citizens of New Orleans and its vicinity, to the rudely constructed log cabin, on the Sabine and Calcasieu, will suggest matter for the deepest reflection. In the short period of ten or fifteen days, can be viewed the moral revolutions of all ages. On a space of three hundred miles can be found human beings from the most civilized to the most savage.[23]

Darby allowed that the "savage"—in this case, the "pastoral creole"—possessed a happiness that might elude "men more highly cultivated"; while he valued the "luxury and learning" of New Orleans, he took a romantic view of the "noble savage" and at times fancied himself another Chateaubriand (whose works he had recently discovered), a solitary wanderer in the heart of the American wilderness. He was convinced that "Nature has everywhere solaced man by some striking benefaction for the hardships of his

22. *Description of Louisiana*, 165–66.
23. *Emigrant's Guide*, 61.

Darby's map of Louisiana, published with *A Geographical Description of the State of Louisiana*, 1816

existence," and after his journey from the Sabine to New Orleans, from the Coushatta Indian village (near present-day Merryville) to the elegance of Royal Street, Darby must have felt a keen sense of hardship at the city's indifference to his discoveries.

The geographer's activities throughout the rest of 1813 remain a matter of some conjecture. Carrying Bartolomé Lafon's 1805 map of Louisiana, he apparently spent several weeks inspecting the marshland and coastal inlets between the Mermentau and Vermilion Rivers. He also sounded the passes at the mouth of the Mississippi in April, 1813. And he evidently participated in some unofficial capacity in government surveys of the Red River between Natchitoches and the northern border of the state, using this occasion to make another tour eastward as far as the Ouachita River.[24] Darby also spent some of the year in Opelousas, where he worked grimly to bring his manuscript closer to completion. As far as possible, he drew upon the knowledge of local denizens who had traveled in Texas and the Missouri Territory; he also read certain portions of the text as public lectures. By mid-summer, 1814, the work was finished at last, and Darby made plans to depart for New York and Philadelphia in search of a publisher.

But that journey was destined to be delayed. While the geographer was rambling through the thickets and swamps of Louisiana, the United States had gone to war against Great Britain; suddenly public events began to impinge on his private designs. Darby had taken up temporary residence in late August with the Bringier family of St. James Parish, whose splendid plantation, White Hall, lay sixty miles up the Mississippi from New Orleans. Louis Bringier, a speculator and adventurer, had prepared a map of the Missouri Territory north of the state of Louisiana, and he planned to travel with Darby to the East to have it engraved. His father, the wealthy émigré Emanuel Marius Pons Bringier, had gone to New Orleans to collect money to finance their trip but returned home in early September with the "prostrating" news that the nation's capital had been sacked and burned by the British on August 25. Faced with the prospect of running a blockade and entering a theater of war, Darby decided to postpone his publish-

24. *Ibid.*, 85.

ing career and wait for some resolution of the conflict. Yet curiosity about the spectacle of war induced him to embark on another journey, an "extensive tour in Florida and Southern Alabama," from mid-September until early December. The entire area had become a scene of military action: British troops had occupied Pensacola while warships hovered off the coast; in Alabama, the Creek Indians had devastated Fort Mims on August 20, 1813, touching off a war which continued to rage in late 1814. In response to both threats, General Andrew Jackson, commander of the Seventh Military District, took Mobile on August 22, 1814, and two months later organized an offensive which led to the capture of Pensacola on November 7. Darby's itinerary during this troubled period remains problematical, but it was obviously influenced by these military circumstances. He seems to have engaged in a more thorough study of Mobile Bay than his earlier visit permitted; this work was apparently carried out in late September and early October in the wake of an abortive British invasion of Mobile Bay on September 16, 1814. He spent some time in the town of Mobile, and it seems reasonable to believe that he may have made Jackson's acquaintance during this excursion, since the general's unfamiliarity with the coastal area made him eager for the sort of geographical insight Darby could provide.

Though Mobile was a seat of military and commercial activity at this juncture, the surveyor took an equal interest in the newly created settlement of Blakeley, situated on the Tensaw River across the bay. This community, the brainchild of Josiah Blakeley, a wealthy New England farmer, had been laid out in city lots during the summer of 1813 (that is, since Darby's 1812 visit), and the vision of a thriving emporium—promulgated by the founder and his associates—apparently captured the geographer's imagination. In *The Emigrant's Guide*, the usually prescient Darby thus delivered the unlikely verdict that in its growing rivalry with Mobile, "the obvious superiority of the position of Blakeley will probably be decisive in its favour." [25] Although the settlement flourished for several years, becoming a shipping center for Alabama cotton, its importance never equaled that of Mobile, and a

25. *Ibid.*, 36.

series of yellow fever epidemics decimated the population so completely that Blakeley became a ghost town about the time of the Civil War.[26]

Other possible sites on Darby's 1814 itinerary are Mount Vernon (which had replaced Fort Stoddert as a militia staging area); Fort St. Stephens (which could be reached by schooner from Mobile Bay via the Tombigbee River); and Pensacola. Since his map of Louisiana subsequently represented the Gulf coastline as far east as Pensacola Bay and included numerous indications of water depth and sand composition at the bottom, it seems clear that Darby personally performed these studies in the aftermath of Jackson's occupation of Pensacola. Indeed, since the British controlled that area until November 7, 1814, Darby could hardly have carried out the work any earlier without arousing the suspicion of enemy naval forces. His arrival shortly after the capture of Pensacola must have given him a vivid sense of the threat posed by the British along the Gulf Coast. An attack on New Orleans seemed inevitable, and alarmed by this prospect, Darby made a hasty return to Louisiana, probably during the last week in November and surely by land, since the King's navy then controlled the coastal waters.

In a state of anxiety, the geographer reached Baton Rouge in early December and there received the stunning news that in his absence Elizabeth Boardman Darby had died on October 23, 1814. Her death must have been utterly unexpected. Yet in one of his last letters to Eliza, written "on the banks of the Mississippi" on August 29, Darby had meditated upon transience and mortality, copying for her edification a passage from Chateaubriand that would later haunt him:

> Full of ardor I launched alone into the stormy ocean of the world
> . . . whose ports and rocks were to me alike unknown. I visited
> places once inhabited by nations that are now no more. I walked
> amongst, and seated myself on the ruins of Greece and Rome—
> countries which still excite strong and interesting recollections.
> Where are the palaces of their kings? Buried in dust. Where their
> mausoleums? Concealed by bramble, or crushed into indistin-

26. See Erwin Craighead, *Mobile: Fact and Tradition* (Mobile, 1930), 140–47.

guishable fragments. Oh! power of Nature and weakness of Man! The tufted briar pierces the hardest marble of the tomb.[27]

Though he could not have foreseen his wife's death, Darby must have felt remorse, even guilt, at having been away on another excursion at the time of her passing. He had acquired new geographical information at an incalculable price.

But his mourning was rudely foreshortened by the news that the British had already launched an invasion upon New Orleans. Settling his affairs in Opelousas and leaving his daughter Frances in the care of fellow surveyor William B. Jackson (who in 1808 had married Maryanne Boardman, Darby's stepdaughter), the geographer hastened to New Orleans to volunteer his services to General Jackson. Even before his arrival, though, the struggle had begun: on December 22, 1814, the British struck through Bayou Bienvenue, taking Jackson by surprise and securing an undefended area. The next night, however, the motley American forces staged a daring attack that halted the enemy advance. Amid preparations for further fighting, Darby arrived in the city on December 27 and proceeded to the home of William Flower, a nephew who had been "shot through both thighs" and taken prisoner in the night attack. At Flower's house he spent the evening discussing war reports and prospects with his brother-in-law, Judge George Mathews, and with Judge Dominick A. Hall, soon to gain notoriety through a legal clash with Andrew Jackson. The following day, the magistrates fled the city, retreating to the security of West Feliciana Parish, while Darby hastened to "the camp of our army" to offer his services. According to a letter published in the *Daily Madisonian* (January 10, 1843), he was engaged as a topographical advisor "with the pay and rank of a Captain in the line," and he set out immediately with a squad of fifteen men to examine the marshes north of the city to determine their susceptibility to an invasion. Darby spent the evening of December 29 "on a hillock within call of the British sentinels," and he reported his findings to Jackson the next day, receiving instructions

27. Darby saved the letter and later published it in the *National Intelligencer* (November 19, 1834). The translation came from Chateaubriand's *Le Génie du Christianisme*, which Darby recommended to his wife as "a work as novel in execution as sublime in the scenes it retraces."

for further reconnoitering. His activities on the day of January 8 remain a matter of conjecture, though in his *Madisonian* letter he did recall seeing "the British veterans recoil from the front of an United States army." As Wilbert S. Brown has pointed out in his military analysis of the campaign, one of Jackson's most pressing concerns as commander of the American forces was "familiarizing himself with the geography and topography of the country around New Orleans."[28] Brown notes that in this undertaking, Jackson was fortunate to have had "the benefit of Darby's counsel" as well as the use of his map of Louisiana.

During the siege of New Orleans, Darby stayed at the Flower house in the Faubourg St. Mary and shared the ordeal with a handful of relatives, gathered to aid in the defense of Louisiana. Among them were two step-sons, Charles and Francis Boardman; a brother-in-law named James Carpenter; George Dewees, the husband of one of his step-daughters; and (after an exchange of prisoners) the wounded William Flower. The men came and went as their duties required, and they accepted unquestioningly the conditions of martial law imposed upon the city by Jackson. Correspondence reveals that by January 27, 1815, the general had appointed Darby "Assistant Engineer pro tem," charging him with the task of surveying all the waterways southeast of New Orleans and blocking or fortifying those which offered access to the British troops still lurking off the coast in ships.[29] The possibility of another invasion was real enough, as the British capture of Fort Bowyer on Mobile Bay demonstrated two weeks later. Appreciative of the enemy's determination, Darby thus made a thorough study of the area under his purview from January 29 to January 31, and on February 1 he reported back to Jackson, advocating the construction of three forts. Only two were authorized by Jackson: one at the junction of Bayou Gentilly and Rivière au Chêne and another on Lake Lery at the mouth of Bayou Terre aux Boeufs. On February 5, the geographer found himself in charge of a contingent of twenty-seven Negro slaves and accompanied by fifteen

28. Wilburt S. Brown, *The Amphibious Campaign for West Florida and Louisiana, 1814–1815* (University, Ala., 1969), 63.
29. Andrew Jackson to William Darby, January 27, 1815, in Andrew Jackson Papers, Manuscript Division, Library of Congress, Washington, D.C.

soldiers who would serve as guards during the construction. He set out for Lake Lery and reached that site on the evening of February 7, reporting to Jackson the next day: "After a painful and troublesome voyage I made good a landing here last night, and commenced the battery here this morning. In addition to the few negroes I brought with me I found 15 Catalans from Terre aux Boeufs. Sick included I will have about 35 workmen generally from 25 to 30 fit for labor."[30] In a week, the fortification had been completed; after the guns were mounted and the garrison stationed, General Edmund P. Gaines inspected and approved the work. At whose suggestion we do not know, but the installation was called Fort Darby.[31]

The second outpost was never built; Darby merely sank a log structure (*cheveaux de frise*) in Rivière au Chêne to obstruct its channel. Then with General Gaines (who succeeded Jackson as American commander at New Orleans) he carried out an inspection tour of various locations still considered vulnerable: Petite Coquille, the Passe Chef Menteur, and the lower portion of Bayou Terre aux Boeufs. But these measures possessed little urgency; rumors had already reached the Southwest that a peace treaty had been signed many weeks earlier. Hence, on February 27, Darby took leave from his duties and returned to Opelousas, two days later writing to a friend: "The Campaign has closed and whether the war continues or not all attempts on New Orleans by his Britanick Majesty's Forces are finished—prodigal as they are of blood—unfeeling as they are respecting human misery or happiness—The venal herd of officers that form the British staff will pause before they again risk their reputation as Soldiers in the American Thermopylae."[32]

The British did not, in fact, strike the Louisiana coast again, but in an effort to maintain vigilance in New Orleans, Jackson kept the city under martial law. When a member of the Louisiana legisla-

30. William Darby to Andrew Jackson, February 8, 1815, in Andrew Jackson Papers.

31. In his *Description of Louisiana*, the geographer proudly referred to Fort Darby, "built during the campaign of 1815, under the direction of the author of this work, by order of general Jackson, to prevent the advance of the enemy" (p. 72). Darby neglected to mention that the fort saw no action and that it was abandoned shortly after construction, but he did mark its location on his Map of Louisiana.

32. William Darby to L. W. (Levin Wailes), February 27, 1815, Louisiana State University Archives, Baton Rouge, La.

ture, Louis Louallier (Darby's neighbor from Opelousas), protested the edict in a French-language newspaper, Jackson ordered his arrest as a traitor. Louallier's attorney petitioned Judge Dominick A. Hall for a writ of habeas corpus, and when Hall complied, Jackson had the judge himself arrested. But a military court acquitted Louallier; realizing that he could not therefore prosecute Hall, Jackson instructed the judge to leave town, but on March 13, news of the peace treaty reached New Orleans, the general rescinded his order of martial law, and Hall returned. A defender of principle, Judge Hall promptly issued a summons for Jackson to appear in his court to face contempt charges. Darby returned to New Orleans toward the end of March and followed the mounting controversy; he was present at the door of Judge Hall's courthouse on Royal Street when Jackson arrived, on March 31, to receive his punishment: a fine of one thousand dollars plus court costs. As Darby reported in his 1843 letter in the *Daily Madisonian*, Hall's punctilious application of justice made him an object of ridicule; Darby remarked to his companions at the moment of sentencing that "the Judge was in a room and on a seat, which his prisoner at the bar saved from being a British barracks."

Darby's acquaintance with Jackson had been brief and impersonal, yet he recognized the fierce authority of the other man, and even after subsequent political differences alienated him from the hero of New Orleans, Darby roundly affirmed in the *Madisonian* letter: "General Andrew Jackson, under heaven, saved Louisiana." Sensible of the renown Jackson had bravely won, Darby called upon him for a testimonial note; with the war at an end, the geographer had revived his scheme to seek a publisher for his study of Louisiana, and he welcomed the general's endorsement:

> Head Quarters, New Orleans, 5th April, 1815
> I have no doubt, whatever, that Mr. Darby's Map of Louisiana is more correct than any which has been published of that country.
> He has certainly taken extraordinary pains to acquire correct information; and so far as my opportunities have enabled me to judge, I am induced to think his delineations very exact.
> Andrew Jackson,
> Maj. Gen. com'g. 7th Mil. Dis.[33]

33. *Description of Louisiana*, 333.

With this and other commendations in his pocket, Darby resigned his position as "Assistant Engineer pro tem" in mid-April to complete arrangements for his voyage to the East.

The prospect of leaving Louisiana created a painful decision: what to do about his young daughter, Frances. The death of Eliza and his own prolonged absence from home had made the girl a virtual orphan. She had resided during his military service at the home of her stepsister, Maryanne Boardman Jackson, who relayed to Darby "solacing" accounts of his "dear little daughter" through the letters of Opelousas Land Office clerk Levin Wailes. Reluctantly, Darby concluded that Frances, a rather sickly child, would receive better care from her stepsister than from an itinerant surveyor, and so he bade her farewell in late April and headed for New Orleans, where on May 7, 1815, accompanied by his friend Louis Bringier, he boarded a ship bound for Philadelphia. He was leaving behind a daughter whom he would never see again (a fate he doubtless anticipated and lamented) and a state to which he would never return. But Louisiana would always remain a subject of intense interest and a focus of poignant personal recollections. To the end of his long life, Darby cherished the memory of his first wife and their daughter and he loved to recount his expeditions through the thinly settled areas of Louisiana. But the voyage in 1815 marked a break with that part of his life; nearly forty years old and toughened by "the accumulated experience gained from near sixteen years of almost incessant motion," he set out to explore new territory—the publishing world of the East.

# 3. Travel and Travail as a Man of Letters

Darby reached Philadelphia in June, 1815, a middle-aged, unemployed surveyor with a manuscript to publish. The past year had been particularly painful—he had lost his wife, relinquished his daughter, and witnessed the horrors of war. True enough, he had fought on the winning side, but the Battle of New Orleans had been horrifying: when spring rains turned the Chalmette battlefield into a bog, decomposed bodies began to emerge from shallow graves; dysentery, malaria, and yellow fever did even greater damage than British bullets.[1] These scenes perhaps repeated themselves in Darby's mind during the weeks at sea. His voyage to Philadelphia was not so much an escape from the past as a logical extension of it, a continuation of his search for a place in the world. After the fiasco at Natchez and ten years of wearisome and unrewarded toil in Louisiana, Darby had at last developed a sense of his vocation as a man of letters. He had come to the East to make known the nature of western lands and western life.

He arrived in the publishing center of Philadelphia with neither the easy optimism of the young writer nor the savvy of a seasoned litterateur, and he quickly discovered two things: few publishing firms were interested in the work of a hitherto unknown author, and the local trade in cartographic publications was controlled

1. Samuel Carter III, *Blaze of Glory: The Fight for New Orleans, 1814–1815* (New York, 1971), 298.

by a single individual—John Melish. Unwisely, as we shall see, Darby made arrangements through Melish in the summer of 1815 to bring forth his study of Louisiana, a work which, as Melish saw immediately, afforded the first complete and accurate delineation of the state. Having obtained a publisher, Darby set out in the autumn of 1815 to revisit the scenes of his boyhood in western Pennsylvania. The pattern of return would become a familiar one: though he lived all but five of his remaining thirty-nine years in the East, he continued to make periodic pilgrimages to the old "Western country" of the Ohio Valley. Interestingly, this same motif forms an integral aspect of his border narratives, many of which are—as we shall see—recounted by a traveler returning to scenes of past experience.

Darby stopped first at Swatara but only for a single night, so anxious was he to see his mother and sister, still living in the western part of the state. We can readily imagine the outpouring of feeling which that reunion inspired and the long hours of conversation devoted to Darby's adventures in the Southwest; he apparently spent several weeks visiting his mother, his "dear sister Nancy," and her husband, Hugh Blair. The journey was not entirely sentimental, however, for during his stay, he decided to undertake a topographical study of Pittsburgh. Indeed, the scheme seems to reflect Darby's characteristic restlessness and passion for information. Actually the enterprise had been suggested by Robert Patterson, a Pittsburgh bookseller, who advanced money to support the study, underwrote the expenses of engraving and printing, and then shared the profits with Darby. Thus, during October and November of 1815, he rambled about the Pittsburgh area, calculating the elevation of rivers and hills, studying the geological strata, and mapping the roads and waterways which converged there. He was particularly fascinated by the commercial and industrial activity; "this city is literally a work-shop," he wrote in *The Emigrant's Guide*, adding that "Pittsburg is by no means a pleasant city to a stranger. The constant volumes of smoke preserve the atmosphere in a continued cloud of coal dust."[2] On balance, however, Darby judged that the area pre-

2. *The Emigrant's Guide,* 258–59.

sented "a smiling picture of active industry and domestic happiness."[3] From the information collected during his autumn sojourn, he produced his *Plan of Pittsburgh and Vicinity* (1817) and the sketch of the city included in *The Emigrant's Guide*.

Darby returned to Philadelphia, probably in late November, to oversee the production of his book on Louisiana and to secure an engraver and publisher for his map of Pittsburgh. He also found time for another pursuit: his courtship with Elizabeth Tanner, whom he married on February 22, 1816, in an evening ceremony at Christ Church, performed by the Right Reverend Dr. White.[4] As in his marriage to Mrs. Boardman, Darby managed to ally himself with a family of means by wedding Elizabeth Tanner, the thirty-year-old sister of the well-known Philadelphia engraver Henry S. Tanner. But we need not impute financial or professional motives for the match; Darby seems consciously to have avoided trading on his brother-in-law's reputation, though Tanner did engrave his map of Pittsburgh, probably in 1816. And there is ample evidence of his genuine esteem for Elizabeth, a "most affectionate companion," who accompanied him on many of his travels in succeeding years.[5]

Two months after his remarriage, the long-delayed study of Louisiana was finally released by John Melish. The book was priced at twelve dollars—quite expensive by 1816 standards—and advertisements described it as "a work of very great importance" produced by "a man of uncommon industry and perseverance, aided by an enthusiastic love for the dissemination of geographical science."[6] But whatever satisfaction Darby derived from this attention must have been short-lived, for within a few weeks he realized how cleverly he had been exploited by Melish. No sooner had his *Description of Louisiana* appeared than Melish brought out a revised version of his own *Map of the United States*, which incorporated—without acknowledgment—the essential features of

---

3. *Ibid.*, 260.
4. *Poulson's American Daily Advertiser*, February 24, 1816.
5. The most poignant testimony is his letter, written shortly after Elizabeth's death, to Lyman C. Draper, September 7, 1847, in Draper Manuscript Collection.
6. The advertisement appears, among other places, in John Melish, *A Geographical Description of the United States* (2nd ed.; Philadelphia, 1816), 169.

Darby's map. In a separate work, *A Geographical Description of the United States* (1816), "intended as an accompaniment to Melish's map," the publisher praised the work of Darby and his companion, Louis Bringier, avowing that "an arrangement" had been made with them, "by which the result of their information was incorporated into this map." As Melish euphemistically expressed it, he was "bringing into view the works of several very meritorious laborers in the vineyard of geography."[7]

But Darby found only sour grapes in Melish's vineyard. His version of the arrangement—substantiated by several acquaintances—presents a very different view:

> When I reached Philadelphia, I found Mr. John Melish in the full possession of the map publishing business, and through him I obtained the publication of my map and statistical volume. But, on what terms? Why, with some other not very moderate terms, to have the privilege of incorporating mine, with other material, on his map of the United States, then in preparation, and soon after published. With these terms I was obliged to comply, and no farther proof need be given to prove that the procedure virtually transferred the real value of my map to Mr. Melish, as his general map contained in a condensed and connected form all my data. But this was not all, nor the worst. Mr. Melish not only secured the profit, but received the credit, and that in a very eminent degree, as the subjoined document will shew.[8]

The "document" was a portion of the treaty with Spain, signed in 1819, which to Darby's everlasting dismay cited Melish as the authority on the Sabine River and the Gulf of Mexico and utilized the Melish map to establish the official boundary line between the United States and Spanish Territory. Darby brooded for years upon this injustice, and in late 1847 he drafted an appeal to Congress for compensation, claiming—no doubt rightly—that his own unaided surveys of the area had proved of substantial benefit to the government. The poverty of his declining years had compelled him to make public the resentment he had borne silently for

7. *Ibid.*, 11.
8. Henry O'Reilly, "Pioneer Geographical Researches," *Dawson's Historical Magazine*, XIII (October, 1867), 225.

so long; after seven years of stubborn effort, Darby received his remuneration—$1,500 granted to him by Congress ten weeks before his death.[9]

Despite all of the complications that it produced, Darby's *Description of Louisiana* made a genuine addition to geographical knowledge, for until his study, the terrain and boundaries of that state were but imperfectly known. An examination of the original map reveals the cartographer's care for accuracy: each river, lake, and bayou is meticulously delineated. The achievement is the more remarkable when we realize that Darby was probably the first man of science to explore some areas of Louisiana. The significance of his work can be inferred from a comment by Samuel H. Lockett, who conducted an important survey of the state after the Civil War:

> A large part of Louisiana was entirely unsettled, and the greater part of it but thinly peopled at the time of Mr. Darby's survey. He tells us that he was the first white man to pass through much of the country he visited. A great deal of country he must have found almost impossible to penetrate. And as there were not then, as now, numerous old settlers, veteran hunters, and hardy lumbermen, who dared these unknown wilds in search of new homes, or game, or denser and more spacious forests of timber trees, he could gain but little information by inquiry. He could know only what he saw with his own eyes. And yet his description of Louisiana is wonderfully correct; and for giving a good general idea of the geography of the state and many of its peculiarities, it is nearly as valuable now as it was when first published more than half a century ago.[10]

Clearly, Darby's "wonderfully correct" study represents his most important original contribution to American geography. The work moreover possessed geological significance; speaking of Darby's investigation of the state's coastal islands, Professor Arthur C. Veatch wrote in 1899: "This is, to the best of our knowledge, the first visit of a man of scientific attainments to any of the

9. See Chapter 5 for a summary of this episode.

10. Samuel H. Lockett, *Louisiana As It Is: A Geographical and Topographical Description of the State*, ed. Lauren C. Post (Baton Rouge, 1969), 4. Lockett's manuscript, prepared for publication in 1873, lay in obscurity for nearly a century following the author's sudden death.

islands. He was a man of keen insight and may justly be regarded as the first to make geological observations of importance in Louisiana."[11] The popular acceptance of Darby's *Description of Louisiana* (which led to a second edition in 1817) established the author's reputation as an authority on the Southwest. Despite his subsequent publications, it would remain the work invariably associated with his name in newspapers and periodicals.

Needless to say, the unfortunate events of 1816 terminated Darby's business relationship with John Melish. The distraught geographer took his map of Pittsburgh (which had been scheduled for printing in April) out of Melish's hands and published it instead through Robert Patterson, the Pittsburgh printer. Still smarting from his initiation into the predatory ways of the publishing world, Darby and his wife left Philadelphia in late summer to visit New York, where in all probability he began to cast about for a new publisher. After attending to practical matters, he and Elizabeth left New York by steamboat on August 20 to tour the Hudson Valley. The excursion provided, as Darby makes plain in *A Tour from the City of New-York, tò Detroit*, a delightful exercise in historical imagination. At the ruins of Fort Putnam, near West Point, he felt himself drawn back to the Revolutionary period, through the mediation of Samuel Johnson: "Looking down from the broken walls of Fort Putnam, Dr. Johnson's Rasselas, came strong to recollection. I could not avoid recalling to imaginary life, the men who once acted on this little but remarkable theatre. I felt a sentiment of awe, amid this now lonely waste, on recalling to mind that here once depended the fate of a new born nation."[12] Darby found the scene a mute reminder of the historical process: though only "fallen fragments of stone" testified to the action at Fort Putnam, the participants had already "taken their respective stations in history."[13] At another site his interest in geological history carried him back to a more remote epoch:

> Environed by the massy and sublime monuments reared by the hand of nature, and enjoying the softened beauty of such an eve-

11. "The Five Islands," in *Geology and Agriculture*, Part V, *A Preliminary Report on the Geology of Louisiana* (1899), Sec. 3, Special Report 3, p. 214.
12. *A Tour from the City of New-York, to Detroit* (New York, 1819), 15.
13. *Ibid.*

ning, I could not repress a retrospection upon the march of time; I could not avoid reflecting that an epoch did exist, when the delightful valley in which I then sat was an expanse of water; that the winding and contracting gorge, through which the Hudson now flows, did not exist, or was the scene of another Niagara; I beheld the lake disappear, the roar of the cataract had ceased, the enormous rocky barriers had yielded to the impetuous flood.[14]

After stops in Poughkeepsie and Kingston—where Darby again relived scenes from the Revolution—the couple reached Albany, from whence they soon made their return to New York and then Philadelphia. The journey was one of several the geographer made during 1816 and 1817; though direct evidence is lacking, comments written in 1818 indicate that he traveled north as far as Boston and south as far as Richmond during this unsettled period.[15]

By the beginning of 1817, Darby had two major projects under way: the second edition of his *Description of Louisiana*, revised in light of new information, and a more comprehensive work, treating the entire West, which he would title *The Emigrant's Guide to the Western and Southwestern States and Territories*. For the latter work, he toiled to prepare an elaborate map of the United States which would in effect compete with Melish's map. But requiring a more immediate source of income, Darby pursued other commercial schemes as well; in March, 1817, he wrote to William Coxe, a member of the New Jersey General Assembly, proposing that he be employed by that body to construct a complete and reliable state map. He took the occasion to criticize the "confined" attitude of many citizens "respecting so important a branch of national improvement, as correct Geographical delineations of their respective States."[16] But this proposal, like others, never came to fruition; hoping to improve his fortunes, Darby and his wife moved to New York City in late spring, 1817.

A few months after he took up residence there, Darby was elected to the New-York Historical Society and thus acquired a circle of informed acquaintances with whom he discussed the findings of

14. *Ibid.*, 12.
15. *Ibid.*, 120.
16. William Darby to William Coxe, March 26, 1817, in Simon Gratz Collection, Manuscript Division, Historical Society of Pennsylvania, Philadelphia, Pa.

his various researches. The society's meetings at the New York Institution in City Hall Park attracted some of the most remarkable men in Gotham: Robert Fulton, Samuel L. Mitchill, James Kirke Paulding, John Jay, David Hosack, and the Honorable De Witt Clinton, then president of the society and governor of the state of New York.[17] With Clinton, the Albany patroon Stephen Van Rensselaer, and others, Darby ingratiated himself as a fellow proponent of "internal improvements"—a political issue of increasing importance during the postwar years. He also became embroiled in his first public literary battle; his adversary, improbably enough, was the amiable and respected Hezekiah Niles, editor of *Niles' Weekly Register* in Baltimore. The affair began when Darby chanced to read an article, published by Niles and written by a certain "Louisiana Planter," which charged that Darby's book on Louisiana was "a much inferior production to his map and not much to be relied on for useful information."[18] The remark aroused such powerful indignation that Darby fired off a letter to the New York *Columbian* on November 12, 1817, blasting the Louisiana planter and editor Niles for the promulgation of incorrect information and incidentally charging three other persons with plagiarism of material from his *Description of Louisiana*. Darby held Niles accountable for "topographical mistakes" in the *Weekly Register* and then moralized on the importance of accuracy: "In every stage of my advance as a writer, however humble may be my attempts, I have constantly endeavored to present facts as they really are in nature. The mischief is incalculable that has been done by high wrought pictures of rapid gain held out to persons moving into the Ohio and Mississippi valleys."[19] Niles responded to Darby's charges almost point by point in the *Weekly Register* (November 22, 1817). Somewhat lamely he admitted that he had not read the offending article before publishing it, but he pointed out that Darby's *Description of Louisiana* had received favorable notices in his journal and demonstrated the unfairness of several accusations by Darby. "There are no classes of men in the world so

17. See the opening chapters of R. W. G. Vail, *Knickerbocker Birthday: A Sesqui-Centennial History of the New-York Historical Society, 1804–1954* (New York, 1954).

18. "Louisiana Planter," *Niles' Weekly Register*, XIII (October 18, 1817), 119.

19. From the New York *Columbian* letter, reprinted in *Niles' Weekly Register*, XIII (November 22, 1817), 198.

jealous of their rights, as inventors and authors," Niles observed, proclaiming his own "sincere devotion to HOLY TRUTH, at all seasons."[20] The feud, apparently a product of Darby's disgust with the conventions of the publishing world, was soon forgotten. But it does illuminate an aspect of Darby's personality—his fierce pride in the accuracy of his work—which occasionally vented itself in petty outbursts. During the next decade, happily, his geographical works received frequent endorsement from the high-minded Niles.

While winter weather still gripped New York in early 1818, the publishing firm of Kirk and Mercein brought forth Darby's *The Emigrant's Guide*, replete with two maps, dozens of statistical tables, and geographical descriptions of all the country west of the Atlantic states. In truth, Darby had borrowed rather heavily from his *Description of Louisiana* in composing the first half of the book, but additional chapters on the Ohio Valley, the Great Lakes, and the progress of canal and road projects made the volume a worthy rival of Melish's *Geographical Description of the United States*. Of particular interest to the general reader is the section "Advice to Emigrants," in which Darby reviewed available geographical and cartographical works (lauding Melish's map of the country) and offered suggestions on matters ranging from land purchases to personal habits. He warned that "it demands excessive labour, severe economy, and exemptions from extraordinary accident, to succeed in a newly settled country," but he concluded that "with caution, temperance, honesty and industry, most men will not only secure competence, but wealth, in any part of the valleys of Ohio and Mississippi."[21] Darby sought to paint a balanced picture of frontier conditions and to avoid "holding up exaggerated prospects of rapid gain," having felt the pangs of failure himself. But he understood the aspirations of his readers and closed his preface by extending "his best wishes for their prosperity."

Written to compete with the handful of western guides and directories then available, *The Emigrant's Guide* attracted wide notice, largely on the strength of Darby's *Description of Louisiana*. The *Analectic Magazine* spoke of Darby as "advantageously known

20. Hezekiah Niles, in *Niles' Weekly Register*, XIII (November 22, 1817), 195, 198.
21. *The Emigrant's Guide*, 293, 296–97.

to the public" through his earlier study and remarked that *The Emigrant's Guide* had been "called into circulation" by "the avidity of the public for all that treats of the new countries." The reviewer praised the "sensible" advice to emigrants and observed that "Mr. Darby gives the best account extant in print" of Mobile and the Alabama territory.[22] The lofty *North American Review* also identified Darby as "the author of a handsome and very valuable map, and statistical account of the state of Louisiana" but gave *The Emigrant's Guide* a mixed review: "We . . . find that he has collected a great deal of valuable information which has never before been published. It is not digested, however, with great skill, and perhaps not always selected with the greatest judgment. But as his life has been spent in the pursuits from which he derived the most important part of his information, relative to the countries which he describes, we ought not to complain that his education has not made him an accomplished scholar."[23] Like the *Analectic*, the *North American Review* found Darby's map of the United States disappointing, but it concluded that the work as a whole bore "marks of intelligence, fidelity and patient industry."

With the publication of *The Emigrant's Guide*, Darby achieved a modest prosperity and furthered his growing reputation as an authority on topographical matters. In the early spring of 1818, Elizabeth gave birth to a daughter, named after her mother, the only child from his second marriage to survive beyond infancy. About the same time, Darby obtained a position with the government surveying crew then engaged in establishing the boundary between the United States and Canada as it ran through the Thousand Islands in the St. Lawrence River. On May 2, armed with letters of introduction from Governor Clinton, he set out for the survey headquarters near Ogdensburg, initially following the same route he and Elizabeth had taken in 1816. The journey provided frequent illustrations of that familiar theme in his writing, the transformation of the wilderness into a "cultivated garden." Passing the "German flats" on the Mohawk River, the scene of "murderous border warfare," Darby noted:

22. *Analectic Magazine*, XI (May, 1818), 368–69.
23. *North American Review*, VII (July, 1818), 269.

Time has changed the drama, the rage of war has subsided, the savages have perished or dwindled to a wretched remnant. Towns, villages, churches, schools, and farm houses, now adorn this once dreary waste. The cultivated mind may shed a tear upon the horrors of the past, but a tear like rain drops in the beams of the sun. A review of the present must be delightful to every generous and feeling heart. It is a picture on which is traced, the most interesting revolution in the moral and physical condition of human nature. There is seen the region, where a few years past, roamed the blood stained savage, and where now dwells in peace and plenty the civilized man.[24]

Quite unconscious of the irony, Darby saw the structures of white civilization—made possible by the elimination of the Indian—as symbols of a triumph over savagery. Inspired by this vision of "peace and plenty" in upper New York State, he reached the surveying camp on May 16, only to find that several persons essential to the project had not yet arrived. After a lengthy delay and much rain, the work began, but it moved forward so slowly that by mid-July, Darby resolved "to quit the business and proceed on a tour to the westward."[25]

His journey brought him a few days later to a place he had longed to visit for years: Niagara Falls. That desire had been quickened, four years earlier, by a reading of Chateaubriand's *Le Génie du Christianisme*, in which the French traveler recounted—in lofty phrases Darby would later emulate—his deeply inspiring visit to Niagara in 1791. With the memory of Chateaubriand's rapturous description shaping his expectations, Darby first approached the cataract (as had the Frenchman) as darkness fell.[26] Only a reading of his account can properly suggest the force of the experience for him—his first hearing of "the deep, long, and awful roar of the cataract," his contemplation of the "darkness, gloom, and indescribable tumult" of the whirlpool beneath, and his fascination with the "sublimity and grandeur" of the rapids above the Falls.[27] Darby regarded the falls as an ambiguous em-

24. *A Tour from the City of New-York, to Detroit*, 52–53.
25. *Ibid.*, 88.
26. See Chapter 2, note 27 herein.
27. *A Tour from the City of New-York, to Detroit*, 160, 161, 163.

blem of change and changelessness; at one point he observed, "My reflections dwelt upon this never ending conflict, this eternal march of the elements, and my very soul shrunk back upon itself. . . . The rock was yielding piecemeal to ruin, fragment after fragment was borne into the terrible chasm beneath; and the very stream that hurried these broken morsels to destruction, was itself a monument of changing power."[28] He contrasted the process of geological change with the course of empire: "Time was when Niagara did not exist, and time will come when it will cease to be! But to these mighty revolutions, the change of empire is as the bursting bubble on the rippling pool, to the overwhelming volume that rolls down the steep of Niagara itself."[29]

After completing a map of the Straits of Niagara, Darby journeyed to Buffalo, where on August 2 he boarded a schooner bound for Detroit. Prone to sea sickness, he spent an excruciating nine days on Lake Erie, as unfavorable winds forced the ship into harbor at Dunkirk in New York and at Fairport, Cleveland, and Sandusky in Ohio. Finally, on August 11, the ship reached Detroit; Darby arrived "considerably fatigued, and very willing to enjoy solid land." He found Detroit still a frontier community and compared it to Natchitoches, in Louisiana:

> Each place occupies the point of contact, between the aboriginal inhabitants of the wilderness, and the civilized people, who are pressing those natives of North America backwards, by the double force of physical and moral weight. In each place, you behold at one glance the extremes of human improvement, costume, and manners. You behold the inhabitants in habiliments that would suit the walks of New-York, Philadelphia, London or Paris, and you also behold the bushy, bare-headed savage, almost in primaeval nudity.[30]

Darby remained in Detroit twelve days, exploring the area and recording observations which comprise a valuable sketch of early settlement there. On August 23, he began the homeward journey, passing through Buffalo and Albany before arriving in New York on September 22.

28. *Ibid.*, 161–62.
29. *Ibid.*, 168.
30. *Ibid.*, 190.

After his reunion with his wife and infant daughter, whom he had not seen in over four months, Darby immersed himself in correspondence and literary work. Gathering together the sixteen travel letters he had written to an unnamed friend in New York, he readied his manuscript for publication, inserting numerous footnotes and explanatory passages. In the midst of this redaction, he happened to read Byron's *Childe Harold's Pilgrimage*,[31] a work which seemed to articulate many of Darby's own sentiments about the natural landscape, and portions of the *Tour*—such as the artificially elegant preface—betray a literary self-consciousness, a studied evocation of the solitary-wanderer tradition. Another shaping influence came in the form of a request from Charles G. Haines to provide information on roads and canals for the New York Corresponding Association for the Promotion of Internal Improvements. Darby's response, subsequently included among his travel letters, expressed his wholehearted support of internal improvements. Convinced that more needed to be said on the subject, the author appended to his manuscript a report from Isaac Briggs (who had appointed him to be a deputy surveyor in Louisiana) on the construction of the Erie Canal and some "General Remarks" calculated to persuade the reader that "every road, bridge, or canal that is formed, of however small extent, contributes to unite society, to promote social and moral intercourse, and to render men more liberal and more happy."[32]

Darby's forthright advocacy of internal improvements plunged him into public debate on what was the central domestic issue during the so-called Era of Good Feelings. His position was perhaps inevitable, given his association with De Witt Clinton, the prime mover in the Erie Canal project. But in supporting the construction of inland roads and waterways, Darby simultaneously endorsed three ideas crucial to much of his subsequent writing. First, he affirmed his belief that westward movement was inherently beneficial and desirable—that is, internal improvements facilitated the noble task of civilizing the backwoods and spreading the blessings of an enlightened society. Secondly, he argued that the development of new inland routes would stimulate trade by

31. *Ibid.*, 164–65n.
32. *Ibid.*, Addenda II, xxxv.

opening new markets, speeding the movement of goods, and reducing the costs of transportation. The belief that commerce held the key to national growth would, during the Jacksonian era, inspire his essays supporting tariffs and the National Bank. Third, Darby underscored the importance of national interests and condemned the opposition of sectionalists as "unwise" and "inexcusable." Such nationalistic sentiment was pervasive in the decade after the War of 1812, but it would remain a basic feature of Darby's newspaper and magazine writing during the long, bitter era of regional divisiveness.

Despite the author's good intentions, Darby's *Tour* was barely noticed by the important reviews in 1819. But it nonetheless remains the most engaging of his major works. As a geographical study, the volume teems with observations on the topography, geology, and botany of upstate New York and the Lake Erie shoreline. Yet unlike Darby's first two works, the scientific details here serve a literary purpose, complementing a travel narrative of considerable intrinsic interest. The geographer's excursion from the burgeoning metropolis of New York to the frontier outpost of Detroit, like his earlier journey from New Orleans to the Sabine River, involved a movement through time as well as space; the traveler again witnessed "man in every stage of his progress, from the palace to the hut." The book's appeal for the modern reader lies, however, not in its symbolic configuration, but in the author's ability to invest particularized descriptions with richly personal connotations. Imaginative or historical associations vivify specific moments of experience; for example, his comments on the geology of several islands in Lake Erie give rise to an idyllic passage: "We were now upon or very near the scene of Perry's battle; the evening was serene and beautiful; our little bark glided smoothly and slowly over the waves, where exactly five years, less one month before, the United States' flag was hoisted over the British ensign. I do not remember to have ever spent an evening at sea with so much pleasure."[33] The traveler's unconcealed delight, here a response to both nature and history, typifies the tone of Darby's *Tour*, the real subject of which seems to be the pleasure of the journey itself.

33. *Ibid.*, 186.

Much of the author's pleasure sprang from an aesthetic response to the loveliness of the landscape; his scientific discernment was coupled with a sensitivity to sublime or picturesque scenes. Thus he painted this characteristic sketch of an upstate village: "This is a romantic village, situated on the slope of the hills, with the Chicktanunda, a large creek foaming over ledges of limestone amongst the buildings, and rushing impetuously down the adjacent declivities toward the Mohawk. The sudden effect of this admixture of houses and cataracts is extremely pleasing and picturesque."[34] Repeatedly the traveler strove to capture the tint of distant mountains, the delightful geometry of plowed fields, or the wild beauty of virgin forests.

More revealing than his aesthetic satisfaction, though, is the sense of astonishment which finds expression at several points in the narrative. Actually, Darby experienced two distinct types of wonder—one inspired by nature, the other by the works of man. The former derived from Darby's discovery that, though he had seen much of the American frontier, certain natural scenes seemed utterly unlike anything he had previously contemplated and evoked genuine excitement. Trying to explain the sensations inspired by unfamiliar phenomena, he said of the view from Utica north toward the St. Lawrence River: "I gazed upon the blue verge before me as if I felt myself entering into a new world. To me this transition was not illusory. Though upon the same planet, and even upon the same continent, the images I now see around me are so different from those I have been for a long period accustomed to behold, that my sensations would not be much more changed if I was transported to another world in reality."[35] Darby's astonishment at the diversity of the North American landscape reminds us that for his generation the frontier was still "open," physically and imaginatively; a "new world" beyond all expectation still awaited the innocent explorer. Later, at Niagara Falls, the geographer would experience the same romantic thrill of the unfamiliar— even as he regarded on a smooth rock the initials of those travelers who had preceded him.

The second type of wonder had even greater significance for

34. *Ibid.*, 45.
35. *Ibid.*, 61–62.

Darby: the amazement inspired by evidence of human progress. Throughout his long career as a man of letters, he consistently portrayed this transformation of the wilderness as an almost magical phenomenon. While admiring the "richly built town" of Canandaigua, New York, Darby learned from a fellow traveler that twenty-nine years earlier an Indian village had stood upon the site; the revelation prompted him to remark: "I could not doubt his information, though there was something in the shortness of the period, when compared with the effects of human labor under my eyes, that seemed almost the effect of magic."[36] Such astonishment recurs frequently in *A Tour*; it constitutes Darby's chief delight as a traveler and accounts for his progressive affirmation in the preface:

> I have for thirty-five years, been a witness to the change of a wilderness into a cultivated garden. I have roamed in forests, and upon the same ground now stand legislative halls, and temples of religion. New states have risen, and are daily rising upon this once dreary waste. I am willing to leave the man unenvied to his enjoyments, who would prefer the barbaric picture now presented by Greece, Asia Minor, Syria, and Palestine, to the glowing canvass whose tints are daily becoming richer and stronger, upon the rivers and hills of North America.[37]

Though capable of savoring the savage wilderness and fond of imagining himself as an American Chateaubriand, Darby grew more appreciative of human improvements than natural beauties. In the passage above, his use of the phrase "dreary waste" to describe the untamed land reveals all. While Chateaubriand (whose *Itinéraire de Paris à Jérusalem* Darby quoted in his preface) wandered through the ruins of the Near East and brooded on the subject of mutability, Darby saw himself a celebrant of the resplendent future: "I would rather indulge my fancy in following the future progress, than in surveying the wreck of human happiness; I would rather see one flourishing village rising from the American wilderness, than behold the ruins of Balbec, Palmyra, and Persepolis."[38] His optimism undiminished by Chateaubriand's somber conclusions, Darby traveled from New York to Detroit in

36. *Ibid.*, 132.
37. *Ibid.*, v.
38. *Ibid.*, iv.

Darby's map of the Old Northwest, indicating his travels, published in *A Tour from the City of New-York, to Detroit*, 1819

1818 to see for himself the "cultivated garden" of the West in which human happiness seemed destined to be secured.[39]

Darby's *Tour* seems to have attracted scant attention in contemporary journals, but some index of its reception can be found in the review by James Kirke Paulding, then emerging as an influential arbiter of taste among the New York literati. Paulding mixed faint praise and fractious blame in delivering his judgment:

> Mr. Darby, in his tour, has materially added to the stock of our information upon the most important topics connected with this state; and it is well he has, for otherwise we should be disposed to be severe upon him for the almost total deficiency of personal incident throughout the volume before us. After reading it through, we know little more of our author, than that he is a very intelligent man, a good geologist, and an excellent geographer. He never admits us sociably and freely to the various interesting interviews with agreeable people, which a traveller enjoys, who rambles in a paradise of sweets, and has nothing to think of but to please and be pleased.

Perhaps misinterpreting Darby's aims as a geographer, Paulding faulted the author for failing to "examine the nature of the human heart," and he further warned: "He that executes a work without touching the springs of human action, may meet with cold approbation, but never will be read with delight." But though dismayed by "the predominance of dry fact," Paulding granted the merit of the study: "Mr. Darby, however, deserves great credit. By giving much information that we had not before, he has laid a foundation, which succeeding tourists may diversify and adorn. The materials are worthy of duration."[40] Paulding's strictures about the aridity of Darby's prose have some justification, but it seems equally apparent that the critic misconstrued the book's object in comparing it unfavorably with conventional travelogues (usually replete with gossip and sentiment) which abounded in books and periodicals of that day.

39. The most influential discussion of the garden metaphor in the mythos of the American West appears in Henry Nash Smith, *Virgin Land*, 138–305. See also Leo Marx, *The Machine in the Garden* (New York, 1967), 73–144.

40. James Kirke Paulding, *American Monthly Magazine and Critical Review*, IV (April, 1819), 402, 411.

With the appearance of his third book, Darby initiated his career as a public lecturer—a venture which for the next three decades would provide occasional financial support for the peripatetic geographer. The lecture hall also provided a forum for the information and ideas he thought essential to the enlightenment of a westward-moving people. Over the years his repertoire expanded and became more diverse, but his favorite subjects remained the history of settlement in America and the natural features of the land itself. Of the papers he wrote in 1819, at least one—the essay on Christopher Columbus—appeared in his later collection, *Lectures on the Discovery of America* (1828). If this work adequately represents the style and substance of his lectures, we must conclude that Darby's lyceum performance was rather pedestrian, long on details and statistics but short on fresh ideas or wit.

Toward the end of 1819, Darby undertook two new literary projects: an edition of Thomas Ewing's *Geography* and a treatise on the history and geography of Florida. Both were essentially mercenary exercises. The New York publisher C. N. Baldwin hired him to rewrite those portions of Ewing's text dealing with American geography; the revised edition appeared in the summer of 1820. In preparing a book on Florida, Darby perhaps hoped to exploit interest generated by its impending entrance into the Union; but the work lacked the freshness of direct observation. Derived entirely from published sources or inferences based on his study of maps, the volume provided little more than a stale rehash of outdated material. The absence of personal references virtually confirms that Darby never traveled beyond Pensacola Bay during his 1814 tour. He gathered his information rather at the American Philosophical Society library in Philadelphia, during the early months of 1820.[41] The finished work, somewhat misleadingly titled *Memoir of Florida*, did not appear in print until March, 1821, published by T. H. Palmer of Philadelphia.

Darby's sojourn in the Quaker city apparently gave him reason to return there in the summer of 1820. One motive may have been an agreement with the Philadelphia publisher of Brewster's Ency-

41. A fact Darby acknowledges in the preface to his *Memoir of Florida* (Philadelphia, 1821).

clopedia to supply geographical articles on a regular basis.[42] He may also have concluded that his lecture series could be offered profitably in the Pennsylvania city. We can be sure of one fact: shortly after he returned to Philadelphia, Darby opened a boarding school at 15 Juliana Street. For the modest fee of five dollars per quarter, he instructed his students in "Reading, Writing, English Grammar, Arithmetic, the use of Logarithms, Plain Trigonometry, Surveying, Navigation, Mensuration, Geography, including the use and construction of Maps and Charts, Dialing, and the Elementary Principles of Astronomy."[43] Exactly how long "Mr. Darby's School" remained in operation is unclear; surely it did not expire for want of a diversified curriculum.

During the early months of 1821, Darby offered occasional public lectures on aspects of climate and geography. Then, on March 23, he initiated a course of lectures on the history of the United States, encompassing the period from the earliest European explorations to the treaty concluding the War of 1812. Darby divided the course into forty individual lectures; tickets for the entire series were priced at five dollars "with a liberal allowance for family tickets." Hoping to play on the cultural pretensions of his audience, he published this appeal for support: "In again presenting myself to the inhabitants of this city for their individual patronage, I too much respect the public before whom I am to appear, to offer one word to enhance the importance of my subject. Geography and History have an exalted rank assigned them amongst the useful sciences and elegant accomplishments which adorn cultivated society."[44] His lecture series ran until early July, after which he devoted the remainder of the summer to editorial work—he had been hired to prepare an edition of Brookes's *Universal Gazetteer*—and to rambles about the countryside of eastern Pennsylvania.

The year 1821 was fraught with griefs and new ambitions for the forty-five-year-old Darby. During that year, he received two saddening letters: one from his first wife's relatives in Louisiana,

42. The arrangement continued at least fifteen years; see the *Notes and Queries* letter, 39.
43. *National Gazette and Literary Register*, November 1, 1820.
44. *Memoir on the Geography, and Natural and Civil History of Florida* (Philadelphia, 1821), n. pag. end material.

informing him of the death of his daughter Frances (she could have been no more than sixteen) and another from his sister Nancy, bearing the news of his mother's death in Montgomery County, Tennessee, where she had moved with her daughter and son-in-law a few years earlier. It is somehow reflective of the rootlessness of Darby's life—and perhaps of frontier experience generally—that circumstances scattered his immediate family across half a continent and that the girl and her grandmother died the same year, having never met. Darby responded to this misfortune in a thoroughly characteristic way, by making a journey, or rather a series of journeys, through the interior of Pennsylvania. These excursions, which occupied much of his time from 1821 to 1824, furnished him with an abundance of new information about the industrial enterprises, agriculture, population growth, road projects, and topographical features of his native state. And wherever he traveled, he encountered reminders of frontier history; on a trip to examine the iron works at Mauch Chunk, for example, Darby passed through Lehighton, where he visited the site of a 1755 massacre and meditated over the common grave of eleven victims, all residents of a Moravian community. He also made a pilgrimage to the ruins of the farm on Mahoning Creek where the family of his friend and mentor, Benjamin Gilbert, had been taken captive by Indians in 1780. Increasingly, Darby found that his studies of the land led back to its people and enriched his sense of the past; a few years later he declared: "Geography derives its highest value as an aid to human history."[45] This awareness convinced him that he needed a new forum—a periodical publication in which he might unite his twin passions, geography and history, and present the fruits of his painstaking research. Like other contemporary men of letters, Darby became possessed by a single dream: to found his own monthly magazine.

Apparently by mid-1822 the geographer had developed definite plans for the journal's contents, prepared a prospectus and a list of potential subscribers, and made preliminary arrangements with a printer. He had settled on a somewhat unwieldy title, which if it

45. *Darby's Monthly Geographical, Historical, and Statistical Repository*, I (September, 1824), 4.

excited little curiosity at least satisfied his urge for correctness: *Darby's Monthly Geographical, Historical, and Statistical Repository*. His advertising campaign began in early September, and the initial response gratified him, for on September 17, he wrote to one prospective patron: "The subscription has been open only a few days, and advances as rapidly as I would dare hope."[46] One of the first to subscribe was Governor De Witt Clinton, the champion of the Erie Canal and an acquaintance from the New-York Historical Society; understandably, Darby dropped Clinton's name freely in letters to other potential supporters.

While he struggled to collect subscribers, the geographer also churned out his edition of Brookes's *Universal Gazetteer*. For this ponderous reference volume, he prepared hundreds of thumbnail sketches of American places and geographical features to augment the foreign articles culled from the 1819 London edition. Noting that he had traveled "extensively" over the West, Northwest, and South—and thus described those regions from personal observation—Darby forthrightly proclaimed in the preface: "It will not be presumptioñ to say, that much is added to the Geography of the United States, by this publication."[47] Though the *Universal Gazetteer* does not lend itself to sustained reading, the work remains a useful index of geographical understanding in 1823. Occasionally entries illustrate the preoccupations of the editor; under the heading of "New York," for example, Darby inserted a glowing, eight-page report on the Erie Canal transmitted to him by Governor Clinton. Nearly the longest entry in the volume, the report expatiates on the "magnificent plans of internal improvement" afoot in New York and the "immense advantages" to be realized therefrom.[48] The geographer apparently completed the material for the gazetteer in early January, 1823, and his publishers, Bennett and Walton, promptly sent advance sheets to Hezekiah Niles, eliciting a puff in the *Weekly Register*: "A new edition of [Brookes's *Gazetteer*], considerably enlarged, and published under charge of that well-known and able lecturer on geography and history, Mr.

46. William Darby to Stephen Van Rensselaer, September 17, 1822, in Simon Gratz Collection, Manuscript Division, Historical Society of Pennsylvania, Philadelphia, Pa.
47. *Brookes' Universal Gazetteer* (Philadelphia, 1823), vii.
48. *Ibid.*, 713–20.

William Darby, is about to appear at Philadelphia. . . . This Gazetteer must be considered as equal to any other that can now be
published—Mr. Darby being devoted to the extension of geographical information—patient, laborious and persevering, as is
shewn in all his works."[49] The cordiality of the notice suggests that
on one of his lecture tours to Baltimore, Darby had mended fences
with the influential Niles after their 1817 tiff.

With the publication of the gazetteer, Darby turned his full attention to the unborn periodical. He had determined that the first
two volumes (twelve issues) would deal principally with the geography and history of Pennsylvania; to acquire complete and accurate information, he set out from Philadelphia with his wife and
daughter on March 2, 1823, on what would be a fifteen-month
tour of Pennsylvania and contiguous areas.[50] His travels took him
into virtually every county in the state and inevitably led him back
to familiar places. In early April he revisited the vicinity of Swatara Creek, called on James Dixon, the son of the old landlord,
and wandered through the churchyard at the Derry Meeting
House—just as he later described the experience in "Ellery Truman and Emily Raymond" (*Saturday Evening Post*, December
7, 1833). Later in the year, he returned to southwestern Pennsylvania and studied the topographical structure of the "peninsula" between the Ohio and Monongahela Rivers. In early 1824,
Darby visited Ohio; he contemplated a tour at least as far south as
Nashville, but an accident at Wheeling, in which some of his
goods were lost and his family nearly drowned in the Ohio, forced
him to alter his plans.[51]

Darby returned to Philadelphia on June 25, 1824, and moved
into a house at No. 2 North Seventh Street, from which place he
conducted business relating to the *Repository*. While preparing the
material for the first issue, he sent off a round of letters in July and
August, seeking new subscribers and encouraging old ones to remit their payment. He had priced his journal at fifty cents an issue,
with the promise of at least sixty-four pages of reading matter and

49. *Niles' Weekly Register*, XXIII (February 8, 1823), 354.
50. *Darby's Repository*, I (September, 1824), 5.
51. William Darby to Patrick H. Darby, February 21, 1825, in Andrew Jackson Papers, Manuscript Division, Library of Congress, Washington, D.C.

applicable maps and charts. Interestingly, in all three departments of the magazine—the geographical, historical, and statistical—he proposed to examine those factors productive of the rise and fall of nations—"the prominent causes of their prosperity or decline." While treating such issues, Darby strove to avoid controversy; his prospectus announced: "News and political discussions are to be utterly excluded."

Nevertheless, the first two issues—the life-span of the *Repository*—provided vigorous support for internal improvement, then a charged political issue. The development of internal transportation routes seemed a noble undertaking, and Darby again expressed little sympathy for those who put sectional loyalty ahead of national concerns: "It is a narrow and contracted view of canal or road creation, in such a country as that of the United States, to consider it of local interest. No canal, or road traversing any state, can have its resulting benefits confined to that particular political section."[52] He inserted into his first issue a tribute to Governor Clinton, which must have made even Clinton blush: "When the rubbish and scaffolding of the human character have alike crumbled to dust, and when the foul passions of his own age have been buried in the grave of the existing generation, then will the name of De Witt Clinton stand in history, stable as the mountains of his own native state, mocking the ravages of time."[53] In case anyone missed the slant of his journal, Darby acknowledged in his second number: "My repository was established in order to aid with all my feeble means, national improvement."[54]

Darby wrote the introduction to his first issue on August 14, 1824, sent the material off to his printer, William Brown, and saw the initial copies of the *Repository* some time in late September, a full two years after he had opened his subscription. He apologized to one patron: "The issue of this commencing No. has been, from various obstacles delayed far beyond my intentions when the first Proposals were presented to the public. I am in hopes that the publication will now advance regularly." But a hint of his critical situation lingers in the closing appeal: "Like most persons strug-

52. *Darby's Repository*, I (September, 1824), 7.
53. *Ibid.*, 53.
54. *Darby's Repository*, I (October, 1824), 136.

gling at once with restricted finances and an expensive undertaking, any advances now made by my patrons will be of the most essential utility, and will be most gratefully received."[55]

He had, to be sure, great hopes for the journal; he intended to include "a neat map of some state" in the opening issue of each volume and to dedicate that volume to a thorough study of the state. "By this means," Darby wrote, "if the work is continued a sufficiently lengthened period, it will contain an Atlas of the U.S."[56] A map of Pennsylvania thus accompanied the September issue, which included a geographical overview of its river systems and terrain, a historical review of the settlement of the New World (probably lifted from his lectures), and a statistical analysis of the preferable routes for canals in Pennsylvania. The October number followed with an essay on the geology of Pennsylvania, remarks on William Penn's colony, and an assortment of statistical arguments for internal improvements. But in a publishing world dominated by lively literary gazettes, monthly reviews, and religious journals, Darby's arid *Repository* probably never had a chance.

Despite the editor's tireless efforts, the journal expired with the October number, the victim of public indifference, rising costs, and, perhaps, Darby's own too-passionate attachment to unvarnished facts and figures. Faced with debts, the geographer opened a map and print shop in his house, still hoping to revive the publication by finding the right kind of financial backing. But his hopes dimmed week by week, and in early 1825, he journeyed alone to the nation's capital to make inquiries and visit acquaintances. In all probability, Darby paid a final visit to his old friend, the Quaker surveyor and mathematician Isaac Briggs, who lay dying in Sandy Spring, Maryland.[57] After watching the course of political events in Washington for several weeks, he also called on his former commander, Andrew Jackson, who had lost his 1824 bid for the presidency to John Quincy Adams—thanks to the notorious "corrupt bargain" struck between Adams and Henry Clay.

55. William Darby to Stephen Van Rensselaer, October 2, 1824, in Simon Gratz Collection, Manuscript Division, Historical Society of Pennsylvania, Philadelphia, Pa.
56. *Ibid.*
57. In his later compendium, *Mnemonika; or, The Tablet of Memory* (Baltimore, 1829), Darby included Briggs among his list of "eminent persons," noting that he died in January, 1825, at Sandy Spring. The author described him as an "eminent mathematician."

Jackson's defeat seemed "disgraceful to the nation" in the geographer's opinion; like many others, Darby regarded the general (at least at this juncture) as the emerging leader of the West, the spokesman for internal improvement. As Darby realized, Jackson's influence remained strong despite his defeat, and mindful of this leverage, he visited the general in mid-February, ostensibly to inquire about his brother Patrick's lawsuit before the Supreme Court and to seek the Indian fighter's advice about "removing to the West."[58] Reminding Jackson of his service as a topographer at New Orleans, Darby also made discreet mention of his need for employment, no doubt hoping that his old commander might find a government sinecure for him.

But he received instead Jackson's advice to emigrate to the Southwest. The interview forced Darby to reconsider his prospects; in a letter to his brother on February 21, 1825, he announced an utterly new scheme:

> If the means were at my disposal I would establish a weekly paper in some one of the populous towns of Kentucky, Ohio, or Tennessee. I cannot but think the present a most propitious moment for such an undertaking. The rapid increase in wealth and population, as well as the growing importance of the west as an integral part of the United States opens a wide, a fertile, and in a great degree an uncultivated field to the Philosopher Statesman and politician. If in my power, no other situation in which I could place myself would be so congenial to my wishes; and I cannot think a 25 years apprenticeship in the study of the real and relative resources of the west a bad school to qualify me for such employment.[59]

If this passage sounds suspiciously like a job application, we must bear in mind that Patrick Darby, a Tennessee lawyer and land speculator of dubious reputation, had purchased the Nashville *Consti-*

58. William Darby to Patrick H. Darby, February 21, 1825, in Andrew Jackson Papers. Jackson was allied with Patrick Darby in a Tennessee controversy over land claims, in which the two sought and received a state court ruling that a property holder must be able to trace his deed back to the original grant or face eviction by another claimant. P. H. D. bought several disputed land titles, and when one occupant contested an eviction, the case led to a suit which came before the Supreme Court of the U.S. in March, 1825. The deed of the occupant was accepted as legal proof of ownership. See Henry Wheaton, *Reports of Cases Argued and Adjudged in the Supreme Court of the United States* (New York, 1825), X, 465–72.

59. William Darby to Patrick H. Darby, February 21, 1825, in Andrew Jackson Papers.

*tutional Advocate* in 1821 to harass his political adversaries.[60] Behind the geographer's fulsome rhetoric about the West lay a covert appeal for employment—a plea evidently ignored by Patrick, who had no intention of turning over his blatantly partisan journal to his high-minded brother.

As he neared his fiftieth birthday, Darby confronted another critical impasse in his "chequered existence." Though the first decade of his "career as an author" had solidified his reputation in the public prints as an authority on the West, he had little else to show for his labors and found himself trapped in a seemingly inescapable pattern of catchpenny drudgery and debt. Mercenary editorial projects had exhausted him, poverty had demoralized him, and so in 1825 Darby began to long once more for the fresh start which the West seemed ever to promise. However, his design of founding a weekly newspaper to till the "uncultivated field" of western resources came to naught. Probably because of his brother's indifference, the geographer abandoned his plans for emigration and instead moved his family from Philadelphia to a snug, rented farm in the Maryland countryside in early 1826. For the next nine years, he lived there in rural retirement, offering occasional lectures, producing more works on geography and history, and launching his career as a writer of magazine narratives. On the farm near Sandy Spring, Darby recovered from the *Repository* fiasco and began to diversify his career as a man of letters.

60. See James F. Hopkins (ed.), *The Papers of Henry Clay* (5 vols.; Lexington, Ky., 1963), III, 460 n2. Also, Thomas Perkins Abernethy, *From Frontier to Plantation in Tennessee: A Study in Frontier Democracy* (Chapel Hill, N.C., 1932), 265–68.

# 4. The Making of a Whig

Haunted for a decade by what he called "that devouring and never ceasing anxiety of mind, which is inseparable from a state of pecuniary embarrassment," Darby retreated from Philadelphia in 1826 to the pastoral simplicity of Montgomery County, Maryland.[1] There, for less than one hundred dollars annually, the fifty-year-old geographer rented a small farm that included a peach orchard, a herd of six cows, and enough timber to provide winter firewood. His experiment in economy—more prosaic, to be sure, than Thoreau's—had by 1833 convinced him that "an immense number of Poor would certainly better their condition by removing from the cities into the country."[2] The benefits of his own relocation seemed obvious, and in 1834 he sketched the substance of this tranquil period: "We live in a fine healthy country, twenty miles north of Washington city, and on a rented farm in a country place, I may repeat, possessing most of the essential advantages without the enormous expense of a city. We keep our own cows, and make their feed from the fields. Our source of living is, however, my pen, which is kept commonly busy."[3] Lectures in Sandy Spring or nearby Baltimore also furnished some income during

1. *National Intelligencer*, November 9, 1833.
2. *Ibid.*
3. *Notes and Queries* letter, 40.

this idyllic epoch, in which Darby at last found a measure of real contentment.

Though some evidence suggests that he made an excursion to North Carolina shortly after the resettlement, Darby devoted his energies mainly to new editorial projects.[4] In February, 1826, Bennett and Walton of Philadelphia engaged him to prepare a revised edition of the *Universal Gazetteer* for the sum of five hundred dollars. The compilation, doubtless a tedious enterprise, occupied him through mid-November, as he added details of 2,900 new towns in America and descriptions of every county in the nation. Significantly, when the volume finally appeared in early 1827, Darby's name had replaced that of the originator, Richard Brookes, in the title; the American had in a sense eclipsed the reputation of his British predecessor. Less scientific hackwork followed the appearance of the gazetteer. In 1828 Darby assembled an edition of *The United States Reader, or Juvenile Instructor*, scrupulously avoiding "every expression which could give an impure idea" and presenting instead to young readers "pictures of mildness, tenderness, kindness to one another, fear of God, and reverence for their parents and instructors."[5] Another commercial endeavor of this period was the compendium of historical and geographical facts arrestingly titled *Mnemonika; or The Tablet of Memory* (1829). The Baltimore firm of Plaskitt and Company also consented to bring out the collection *Lectures on the Discovery of America* in 1828, making available in print the very disquisitions he had been presenting from the platform since 1818.

Amid such labors, Darby managed to produce one work of substance and interest, the encyclopedic *View of the United States, Historical, Geographical, and Statistical*, which utilized most of the material originally prepared for the short-lived *Repository*. In a single volume, Darby attempted to condense his knowledge of the topography, climate, and history of America; like many of his tomes, the *View* contains much pedestrian commentary and a

---

4. Darby mentions his travels through the pine forests of North Carolina in *View of the United States, Historical, Geographical, and Statistical* (Philadelphia, 1828), 5. I have not located any other evidence of this journey.

5. "Preface," *United States Reader, or Juvenile Instructor* (2nd ed.; Baltimore, 1830), iii–iv. Darby's preface is dated November, 1828.

surfeit of sheer detail. But one also discovers, here and there, several ingenious and engaging theories on geological evolution, meteorology, and population growth. He developed, for example, a concept of "isothermal parallels" to explain similarities and differences in weather patterns, and he devised a ratio to project demographic trends in the United States through the year 1940.[6] Though many of Darby's theories now seem quaintly naïve, we may still appreciate his determination to identify the laws and principles which would explain natural and social phenomena. Contemporary reviewers greeted Darby's *View* respectfully; with advance sheets at his disposal, Hezekiah Niles plumped the work in his *Weekly Register*: "We feel confident that when it shall be published, it will sustain the good opinion that we hold of it, formed after a careful examination of certain parts of its contents, with which we ourselves profess to have some knowledge—and a belief built on our personal acquaintance with Mr. D. that he would toil for a whole month to correct a single error, however trifling it might appear to other men with differently constituted minds."[7] The *American Quarterly Review* hailed the book as a valuable digest, "a complete view of the physical geography of this country, no where else to be found," but after praising Darby's "manly language," found some fault with his "stiff, and somewhat affected" style.[8] Three years later, a reviewer in the same journal carried the criticism further, deeming the work too theoretical for the masses: "The 'Views' of Mr. Darby abound in general information, of sufficient interest, and might be called the abstract and theoretical geography of the country. But they are much better suited to scholars and men of science than to common readers."[9] For the very reasons that it would now interest the student of American geography, the *View* fell into eventual disfavor with the reading public of Darby's time.

As if such ponderous works did not provide sufficient employment, Darby embarked on an entirely new species of writing in 1829—a series of frontier narratives, sketches, and essays prepared

6. *View of the United States*, 350–51, 438–42.
7. *Niles' Weekly Register*, XXXI (October 18, 1828), 114.
8. *American Quarterly Review*, V (March, 1829), 151.
9. *Ibid.*, IX (June, 1832), 287.

expressly for Samuel C. Atkinson's popular *Saturday Evening Post* and its monthly counterpart, the *Casket*. Adopting the pseudonym "Mark Bancroft," the geographer wrote dozens of magazine pieces between 1829 and 1836 and earned a modest reputation as a teller of western tales. These periodical effusions afforded him imaginative freedom; opening the rich store of his experience and memory to the reader, Darby sought to articulate the meaning of the American West as it embodied itself in actual personalities and incidents. Though he occasionally embellished his narratives with a sentimental theme—he wrote for a largely female audience—the author maintained his reverence for fact and persistently regarded himself as the historian of the old border country. In several respects, Darby's stint as a magazinist seems the most personally revealing phase of his long career, but that discussion must be deferred for a time, so that we may complete our survey of his auctorial and intellectual pursuits during the clamorous Jacksonian era.

The political ascendancy of Jackson in 1828 in fact produced a major turn in Darby's professional life. After observing Jackson's rise to power for several years, the geographer reached some disturbing conclusions about his old commander: the general seemed too willing to compromise himself to solidify political support and had evidently forsaken the great western cause—internal improvement—to mollify influential easterners. Aggrieved by such developments, Darby took the surprising action of lodging a personal protest in the columns of a Baltimore newspaper, accusing Jackson of capitulating to "friends" who were dedicated to his "utter destruction."[10] Identifying himself as one who had once been a "sincere advocate" of Jackson's character and principles, he spelled out three shameful promises that the general had purportedly made: to obstruct all programs of internal improvement; to eliminate from government all supporters of "an American system of internal prosperity"; and to appoint cabinet advisors for whom he would become the "blind instrument" of their will. More a critique of Jackson's "friends" than the candidate himself, Darby's

10. Darby reprinted the letter in full in one of his later attacks on Jackson in the *National Intelligencer*, May 30, 1836. The Baltimore newspaper was not identified by Darby, nor have I located the original appearance of the letter.

letter nevertheless marked him as a political enemy. Though it had no discernible effect on the Jackson camp, the letter seems an interesting gesture in three respects. First, it expresses a blank disbelief, shared by many voters in 1828, at the general's seeming vacillation. Secondly, its blunt, admonitory tone and pointedly personal observations suggest that Darby published the letter with the clear assumption that this complaint by a private citizen would surely command the attention of the presidential candidate—an assumption indicative of an earlier, more intimate political milieu than our own. Finally, the letter marks the beginning of the author's involvement in partisan debate; after his stand against Jackson, Darby began to ally himself with the Adams-Clay forces (the National Republicans), and in 1830 he formalized his opposition to the president by becoming a correspondent to the Washington *National Intelligencer*, the recognized organ of the Republican (Whig) party.

Under such pseudonyms as "Tacitus," "Agricola," and "A Traveller," Darby began to publish learned commentaries and letters in what was perhaps the most widely read newspaper of his day. His earliest known contribution to the *Intelligencer* (October 12, 1830), a homily on the lessons of European history entitled "To the People of the United States," anticipates the method he would employ in much of his later journalism: an explanation of present conditions through a scholarly examination of the past. Through occasional columns, the sententious Darby contemplated some of the leading topics of the day: the threat to the balance of power in government posed by the presidential veto; the need for a sound national bank and a system of protective tariffs; the urgency of continued internal improvement; the economic argument for slavery (the former cotton planter took a sympathetic view); and the vexed question of statehood for Texas. He also discussed, in a lengthy series of essays, the geography and history of "The Northern Nations of Europe," and for a time he translated articles on European life and literature from *Le Courier Français*. In later years, he wrote increasingly about the growth of America's population, the transformation of the West, and the passing of the old pioneers.

Darby's relationship with editors Joseph Gales, Jr., and William

W. Seaton was an enduring one, based upon both professional respect and political compatibility. From various editorial notes, it seems clear that Gales and Seaton considered him a valued correspondent—an eminent authority on geographical and historical matters as well as a man of absolute integrity.[11] Darby's quarter-century association with the *Intelligencer*, terminated only by his death in 1854, had several tangible effects on his career as a man of letters. By providing him with a steady income, it relieved him of the need for hackwork and enabled him to explore the entire range of his intellectual interests. The relationship also strengthened his commitment to the political and economic views which in 1834 became the philosophy of the newly formed Whig party. Though he detested partisan wrangles, a sense of principle compelled him, as it had in 1828, to speak his mind on national events, and he assumed the role ostensibly as an expression of intellectual duty. Finally, through his connection with the *Intelligencer*, the self-educated Darby began in 1830 to conceive of himself as a scholar and philosopher, and his reading broadened to match his new sense of avocation. Eight years later, the transformation would be complete; he would become Professor Darby.

In the same year that he became a correspondent for the *Intelligencer*, Darby embarked on one other notable venture. On February 19, 1830, he received a letter from Edward Hopkins of Hartford, Connecticut, proposing that Darby collaborate with Theodore Dwight, Jr., to prepare a "Gazetteer of the United States."[12] Hopkins shrewdly recognized the commercial advantage of enlisting two well-known geographers; Dwight had consented to furnish all articles concerning the Northeast, including New York and New Jersey, while Darby was assigned responsibility for the rest of the country. According to the latter's account, the project occupied his time "with little intermission" for over two years; but the job paid well ($800) and Darby was able to gather much of the requested material by culling his 1827 *Universal Gazetteer*. He collected new details, however, by placing requests for local information in newspapers across the South and

11. This estimation, reflected in numerous editorial notes over a twenty-five year period, informs their obituary for Darby in the *National Intelligencer*, October 10, 1854.
12. Darby outlined the history of this project in the *National Intelligencer*, May 10, 1834.

West. The completed volume, *A New Gazetteer of the United States of America*, appeared late in 1832, receiving immediate recognition as the most authoritative work of its kind then available. Naturally, the friends of Darby and Dwight contributed accolades; Samuel C. Atkinson included a typical puff in the *Saturday Evening Post* (November 16, 1833): "The name of William Darby, one of the editors, is so well known in this city as that of a man peculiarly well qualified for the task, that it is almost sufficient merely to draw the attention of our citizens to the work to ensure their subscription." Not all notices proved so cordial, however.

When a second edition of the gazetteer appeared in early 1834, one reviewer had the temerity to challenge its accuracy and brought down upon himself the heated indignation of Darby. Writing in *Waldie's Circulating Library* (also called the *Journal of Belles Lettres*), the critic labeled the volume a "catch-penny" and listed a string of trivial errors. Darby considered this a piece of villainy; in a letter to the *Intelligencer* (May 10, 1834), he denounced the critic, recounted his own considerable experience as a geographer, and adduced testimonial letters from five respected acquaintances. He also published a bristling reply in the *Post* (May 17, 1834) in which he righteously insisted:

> Now I can say to the public, and set successful denial at defiance, that for *thirty years*, I have been sedulously engaged in pursuit of correct and important Geographical information; that in that long period, when compared with individual human life, have actually acquired, and by publication, have added more to the science of Geography, than has been done in the same time, by any other unaided individual in existence. This is bold language, and under other circumstances, would be inexcusable language, but its truth, and the atrocious attempt made to deprive my family and myself of the labours of a life, plead full justification.

Darby explained the division of responsibilities behind the work and defended his methods of gathering information. He then closed with a witticism, noting that every reader of a gazetteer finds the account of his own particular locality too scanty: "My critic of the Journal, appears to have had his gall stirred up from guardian care for Burlington and Bristol; the former particularly, 'where,' says the critic, 'many besides ourselves will remember having Latin beat into them.' I have already shown, that for that part of the Gazetteer, I am not accountable, and therefore, on Mr. Dwight devolves the fearful responsibility of neglecting *the very place* where the editor of the Belles Lettres Journal had Latin beat into him."

Darby's boast that he had "added more to the science of Geography" in his lifetime than "any other unaided individual in existence" surely reveals more about the man himself than the state of scientific knowledge in 1834. Undoubtedly he *had* compiled and published more geographical data than any American since Jedidiah Morse, but his was by no means the most significant theoretical work, and numerous writers had conducted more extensive explorations of the North American continent. What Darby meant, it would seem, is that the incessant accretion of information, to which he had devoted his life's energies, entitled him to more respect than he had received from the *Waldie's* reviewer. But the remark also indicates the presence of a healthy ego in a man customarily described by associates as "modest," "unassuming," and "unostentatious." As this episode reveals, Darby was capable of vigorous self-applause when his reputation was called into question.

About the same time that Darby scrimmaged with the *Waldie's* critic, he was establishing a unique friendship, wholly through correspondence, with Dr. Matthew L. Dixon of Winchester, Tennessee. In December, 1833, both the *Post* and the *Casket* had featured Darby's "Ellery Truman and Emily Raymond, or The Soldier's Tale," a narrative set on Swatara Creek at the time of the Revolution, which contained important references to John Roan, John Dixon, Robert Dixon, and Lindley Murray. Those men happened to be the great-uncle, grandfather, uncle, and cousin, respectively, of M. L. Dixon, who immediately wrote to Samuel C. Atkinson, seeking the identity of the author so intimately acquainted with his family. Darby received this inquiry through Atkinson in mid-February and shortly mailed to Dixon an account of his association with the family and the Derry settlement.[13] A subsequent letter from Dixon prompted the geographer's lengthy

13. *Dixon's Ford* letter.

autobiographical epistle of April 18, which recounted his "chequered existence" from the Pennsylvania frontier and his years at Natchez to his service under General Jackson and his career as a writer and editor. Somewhat immodestly Darby quoted in full his letter of commendation from Jackson (despite his political differences with the president) and a note in the *Intelligencer* describing him as "a person of great intelligence . . . better versed in History 'than any other individual in the Union.'" He endeavored to portray himself as the self-educated man of letters: "My reading has been desultory, I confess, and far indeed from that of many, but it has been beyond what is commonly attempted by persons of straitened means, and not professionally engaged. You see, I am laying my heart naked to you . . . to demonstrate that a tolerable education is within the reach of every free white in the United States." [14] He added the incidental note that he had read in English or French "every one of the most eminent classics."

The geographer's correspondence with Dixon and his mother, Mrs. Anna Cochran Dixon, continued for sixteen years and apparently provided Darby with new insight about his past, reawakening his earliest memories. He thought so highly of Dixon's communications that he forwarded one to Gales and Seaton for publication in the *Intelligencer* (May 30, 1835) as a historical document. The epistolary friendship developed, coincidentally, at the very time Darby began a "regular engagement" to supply border narratives, and Dixon (who was both a physician and bank clerk in Tennessee) seems to have been eager to offer ideas and materials for such tales. At least one "Mark Bancroft" tale received an impetus from the correspondence, for Darby wrote to Dixon on April 18, 1834: "The incident of the capture and recapture of your mother-in-law and Boone's daughter has been long since fixed in my eye as a chosen subject, and this added to the extraordinary fact of my having connected in the same tale the families of both your parents, gives true interest to the series of circumstances." [15] Two months later, the *Post* featured "Cyrus Lindslay and Ella Moore," a tale which linked the experiences of Dixon's relatives with the career of that preeminent hunter-warrior, Daniel Boone.

14. *Notes and Queries* letter, 39.
15. *Ibid.*

Dixon's letters may have had one other effect: they apparently re-kindled Darby's desire to travel. He yearned to meet Mrs. Anna Dixon and to share recollections of the "soul-pleasing history of the society along Swatara," but at the age of fifty-eight, other considerations obtruded: "Sometimes I think of making a *Grand Tour* . . . but advancing age and a family, small as it is, but who must go wheresoever I go, as also an old Lady of the name of Prudence, all say, 'sit still, old man, you have already paid away your best days in traveling.'"[16] He could scarcely have guessed that he would make several more journeys to the West before age and infirmity at last compelled him to "sit still."

As we have noted, Darby's years in Sandy Spring were marked by productive contentment. Yet late in 1834, for reasons perhaps related to a change in property ownership, he was compelled to leave the Maryland farm, that "cherished place of long residence," to seek another country dwelling.[17] This he discovered in West Chester, Pennsylvania, an area he considered "one of the most delightful spots in our chequered country." In 1835 he sketched the attractions of the surrounding farmland: "Spreading far to right and left, sweeps a landscape, over which, as on an immense painting, are commingled farm houses, fields of luxuriant grain, meadows of most abundant herbage, orchards, clumps of trees, and roads like great lineaments on the face of nature. To this seductive scene, life is imparted by flocks and herds, the busy ploughman, the reaper, and the many other avocations of human life, but above all the trains of cars literally flying along the Pennsylvania rail road."[18]

During the winter and early spring, Darby seems to have worked primarily on essays and narratives for Atkinson's periodicals; in June, 1835, he resumed his regular contributions to the *Intelligencer*. His yearning for travel also returned in 1835, and in June he made a two-week tour through neighboring areas of Delaware and New Jersey, primarily "to examine the appearance of crops." In a letter to the *Intelligencer*, written June 25, 1835, he

16. *Ibid.*, 35.
17. Darby referred indirectly to this move in a letter to the *National Intelligencer*, November 19, 1834.
18. "Summer Travelling," *Casket* (October, 1835), 588.

spoke of plans for an "extended ramble" through Pennsylvania, New York, and Connecticut, but in late summer he set out instead for "the Western country" to visit once more the places where he had spent his youth and early manhood.

Passing through Pittsburgh, Wheeling, and the intervening villages and towns, Darby was surprised by the rising prosperity of the old border country. Of this experience he observed, "We may approach a place we have never seen before with lively curiosity, but there is a thrilling interest, an indescribable pleasure attending our return to a place in which we have resided years past, far more exquisite than mere curiosity can ever bestow."[19] This delight illustrates again the basic turn of Darby's mind. He was of course intensely responsive to historical associations, but his was no search for a static past. Rather, his "indescribable pleasure" seems to have derived largely from his appreciation of change; the same sensibility which compelled him to support internal improvement enabled him to take pleasure in the evidences of human progress. Wherever his travels took him, Darby continued to prefer the "cultivated gàrden" to the savage wilderness; the rugged land along the Juniata River once prompted him to remark:

> Such scenery, grand, awful, but savage, arrests, indeed, enchains, the attention at first—but I could perceive it on others, and felt very strongly on myself, that there was a want which was increased as the heart sighed for human habitation. The feelings of man, as a social being, cannot, more than a few brief moments, enjoy pleasure from landscapes on which the residences of his fellow-men are frowned away forever. This happy force of humanity, which compels us to mingle our sympathies with those of our species, is called into action, and radiates the countenance of the pilgrim over this earth.[20]

At Wheeling, in the summer of 1835, Darby recalled the history of the place and his acquaintance with its heroes, but he took equal pleasure in the town's emergence as "one of the great emporia of 'the West.'"[21]

19. Darby's travel letter from Wheeling, in *Casket* (September, 1835), 494. The date of composition, given as June 2, 1835, is almost certainly a misprint, since his contributions to the *National Intelligencer* place him in West Chester, Pa., in early June.

20. *National Intelligencer*, August 24, 1836.

21. The *Casket* travel letter from Wheeling, 494.

Darby returned to Chester County in early September, 1835, and took up residence at Kennett Square, a village near the Delaware border. There he produced a plethora of essays and letters for the *Intelligencer*, averaging one major article per week through the end of the year. Consistent with his passion for European literature and political history, he spent much of his time perusing French newspapers, such as *Le Courier Français*, from which he excerpted numerous articles for translation and commentary. By February, 1836, he had moved to New London Crossroads (also in Chester County), and through the spring months he discussed in the *Intelligencer* three political concerns which then absorbed his attention: slavery, internal improvement, and presidential power. Concerning chattel slavery, Darby had argued the necessity of the system, described its benefits for master and slave, and insisted that the "abolition scheme must be destroyed" in an essay on September 14, 1835. In a subsequent treatise (January 23, 1836), he endeavored to demonstrate the sanguine effects of servitude with demographic tables showing the slave to be more prolific than the freedman. Clearly, he perceived the issue as an economic rather than a moral question. The theme of internal improvement emerged in an article of February 25, 1836, supporting the construction of the Atchafalaya Railroad in Louisiana. As in an earlier discussion of the subject (June 6, 1835), Darby made use of his notes on the state to sketch the country to be traversed by the line, opining that "the proposed railroad would operate as if by magic, and disclose to the People of the United States one of the most interesting sections of their great domain."

Darby's principal political concern of the mid-thirties, however, was the concentration of power in the office of the president. In an *Intelligencer* essay of July 30, 1834, published originally in 1826 (before the Jackson era), the correspondent pointed out that the veto power created a chief executive "whose legislative wisdom and foresight is, by the Constitution, considered as superior to any majority less than two-thirds of both Houses of Congress." In a series of letters and essays between 1834 and 1836, Darby hammered away at the issue of presidential power and called for revision of the Constitution. In a short note (August 27, 1834), for example, he observed that one purpose of the Constitution was to establish the Senate as "a counterpoise to Presidential power," and he added the rueful query, "Has not the intention been a failure?" On June 27, 1835, he described the presidency as "an office in itself utterly at variance with all our pretended first principles . . . clothed with absolute authority over the legislation of the country." Capitalizing on Whig sentiment against "King Andrew," he altered the terms of the attack on January 30, 1836, noting that "we have created an officer of the highest dignity . . . and have clothed him with regal power, and called him, not by the old Saxon title KING, but President." Appealing to logic rather than party feeling, though, he insisted on March 4, 1836: "The same man who is to see the laws put into execution should, on all the rules of common reason, have naught to do with their enactment." In a subsequent essay (March 21, 1836) he complained, "This Constitution has given the control of all the great departments and of the finances to one magistrate." What is remarkable about Darby's crusade to limit presidential power is his insistence that the so-called usurpation of the office by Jackson was a natural development of authority granted by the Constitution. Though a confirmed Whig, the correspondent declined to assail Jackson himself and maintained a high-minded tone in the midst of political uproar. But his equanimity was shortly destroyed by a bizarre newspaper controversy.

In May, 1836, editors and public officials in the nation's capital debated the thorny problem of statehood for Texas. While Jackson worried about maintaining relations with Mexico, others raised new questions about the extent of Texas and the location of its common border with Louisiana. Interested in the discussion, Darby journeyed to Washington from New London Crossroads to lend his expert opinion. He consulted briefly with the president himself, who, to the geographer's amazement, advanced the view that there existed more than one Sabine River between Texas and Louisiana.[22] Through essays in the *Intelligencer* (May 19, May 21, 1836), Darby endeavored to put such rumors to rest, but by questioning Jackson's grasp of the issue, he drew hostile fire from the Democratic *Daily Globe* (May 20, 1836), which impugned his in-

22. Darby refers to his meeting with President Jackson in the *National Intelligencer*, May 30, 1836.

tegrity and credibility. Suddenly the boundary question became incidental to the question of reputation; determined to have the last word, Darby answered in the *Intelligencer* (May 30, 1836) with an open letter to Jackson, implying that the president had sponsored the abuse, the perverseness of which might be seen by comparing the *Globe*'s "brutal personal attack" with the letter of commendation Jackson had originally written for Darby in Louisiana in 1815. Charging that the president had gained his office by losing his scruples, Darby sternly remarked: "Unjust exertions of power, direct or indirect, always have and ever must inflict ten thousand times more deadly injury to the oppressor than to the oppressed." This was of course wishful thinking on Darby's part; by 1836 Jackson had become inured to such lofty warnings and probably took little notice of the assault, though he may indeed have inspired the initial, defamatory notice in the *Globe*. Such were the methods of political warfare during the 1830s.

As if to escape further treachery at the hands of the "Jackson men," Darby accepted a new assignment from Samuel C. Atkinson in July: to make a tour of Pennsylvania and Ohio, providing travel sketches for the *Post*, the *Casket*, and Atkinson's latest venture, the *National Atlas*, whose "geographical portion" had been placed "under the care of Mr. Darby."[23] On July 23, the correspondent left Philadelphia with his family, bound for Pittsburgh and the western country. Though approaching his sixty-first birthday, Darby felt again the old pleasures of the journey; interim reports from Middletown and Hollidaysburg, as well as his remarks from Pittsburgh, dated August 4, 1836, all suggest a lively interest in the changing aspect of the Pennsylvania countryside. The author was delighted with internal improvements in the western part of the state and took the occasion to advocate construction of a canal from Pittsburgh to Lake Erie.[24] By river steamboat, he visited the towns of Beaver and New Brighton, filing reports to Atkinson and to William D. Wilson, editor of the Pittsburgh *Advocate*. To the latter, he promised further observations on his tour, expressing once more the spirit of the astonished traveler: "You will hear from me at the different points of my tour, as I love, with

23. Editorial column, *National Atlas*, July 31, 1836.
24. *Saturday Evening Post*, August 27, 1836.

the feelings of an old man, to be communicative of the delight I myself enjoy in traversing hills where new-born evidences of civilization present themselves at every step."[25]

As it happened, Darby foreshortened his ramble to accept a position as co-editor of the *Advocate*, on or about August 20, 1836. Abruptly severing his regular affiliations with Atkinson and the *Intelligencer*, he took immediate charge of the newspaper's "literary and political departments," an assignment which plunged him back into partisan debate. Reflecting the political convictions of editor Wilson, the *Advocate* supported the Whig cause with absolute vehemence, and a few days after assuming his new duties, Darby fired his first long-distance salvo, an endorsement of the candidacy of William Henry Harrison, which simultaneously denounced Jackson's domestic policies, his "Kitchen Cabinet," and his selection of Martin Van Buren as heir to the throne.[26] Though the absence of complete files of the *Advocate* for 1836 makes it impossible to trace Darby's participation in the election battle that autumn, we can assume that this position involved him more deeply in partisan journalism than any other in his career. Yet his connection with the *Advocate* lasted only a year; perhaps as a consequence of financial hardship during the Panic of 1837, Wilson sold his interest in the newspaper and Darby found himself again without steady employment. As co-editor of the *Advocate*, he had been able to realize, at least partially, his old desire to publish a newspaper in the West, but hard times forced him to look elsewhere for his livelihood.

For exactly the same reasons which governed his removal to Sandy Spring eleven years earlier, Darby took up residence in the fall of 1837 on a rented farm in Canonsburg, a village twenty miles southwest of Pittsburgh. Tax records for this period indicate that the family kept a single milk cow as they struggled to realize the benefits of rural economy. Other motives perhaps figured in the relocation: in moving to Canonsburg, Darby was in a sense

25. Reprinted from the Pittsburgh *Advocate* in the *National Intelligencer*, August 24, 1836.

26. Reprinted from the Pittsburgh *Advocate* in the *National Intelligencer*, August 25, 1836. Signed "A Native Citizen of Pennsylvania," this letter surely came from the pen of Darby, the new editor of the *Advocate*'s "literary and political departments."

coming home, since his father's old farm on Chartiers Creek lay only a few miles to the south. Moreover, the village was the seat of Jefferson College (founded by the venerable circuit minister John McMillan), an institution of higher learning which might well employ a lecturer of Darby's experience. The hope became a reality: in March, 1838, he was elected to the faculty of Jefferson College as professor of history, geography, and astronomy, though in fact he served more precisely in the role of visiting lecturer.[27] In the five terms of his nominal connection with the school, Darby conducted classes during only three of them. Wholly lacking in formal academic training, he had secured his temporary appointment on the strength of his general reputation as a scientist and historian, and he must have relished the association with his learned colleagues. Already past sixty when he began his professorial career, he seems to have been a vigorous lecturer; one former student, writing for a class reunion fifty years later, recalled Darby as a man of "rough exterior, but of real learning in his department." The student added, "My own recollections of him are indissolubly associated with 'Opelousas and Attakapas,' which he had personally explored, and loved to tell about."[28] Some early records list Darby as the possessor of a Master of Arts degree, probably as a conventional mark of respect for his faculty status. During his first year at the college, he was elected to honorary membership in the prestigious Franklin Literary Society, placing him in the company of such contemporaries as Mathew Carey, N. P. Willis, Daniel Webster, Edgar Allan Poe, and Thomas Jefferson.[29]

Like many others, Darby found the position of college professor honorable but ill-paying and he began to cast about for other sources of income in 1838. In the summer, perhaps during a

27. *Biographical and Historical Catalogue of Washington and Jefferson College* (Cincinnati, 1889), 12. Darby's tenure is here listed from March, 1838, until 1841. Actually, he resigned in late 1840, according to a note in the *National Intelligencer*, March 16, 1841. Correspondence indicates that he taught during the summer term, 1838, the winter term, 1838/39, and the summer term, 1840.

28. H. A. Brown, "Class Oration: Reunion Proceedings of the Class of 1840," *The Annual of Washington and Jefferson College for 1890* (Washington, D.C., 1891), 53.

29. *Catalogue of the Franklin Literary Society, Jefferson College, from 1797 to June, 1853* (Pittsburgh, 1853), 5.

recess in classes, he conducted a brief tour in Ohio with his wife, lecturing on geography and history in several small towns.[30] Eager for editorial work, he wrote in late 1838 to David Christy of Oxford, Ohio, the publisher of the defunct *Historical Family Library*, suggesting works worthy of reprinting, should the journal be revived.[31] In exchange for his help in obtaining subscriptions, Darby asked Christy to promote his forthcoming lecture tour in Ohio. He also made known his willingness to edit or translate works for the *Library*, but he metaphorically reminded Christy of his pecuniary difficulties: "If every circumstance can be made to fit, I know of no other employment I would prefer to either compiling or translating for The Library; but as few writers, even good ones, or indeed eminent ones, have much *Honey* in their *Hives*, and as I am not one of the few, and can't either live or keep my family alive without honey, I am compelled to fly from flower to flower."[32] The connection with Christy never materialized—nor did Christy's refurbished *Library*.

In such circumstances, Darby also appealed to his old friends Gales and Seaton of the *Intelligencer*, who welcomed the resumption of the geographer's contributions. Thus began his most prolific period as a correspondent. During 1839, he published more than sixty articles in the newspaper: over the pseudonym Tacitus he supplied a lengthy (and ultimately controversial) series of essays on "The Northern Nations of Europe," and over the pen names A Traveller or A Traveller in the West, he furnished historical and geographical accounts of the Upper Ohio Valley and "the Great West," along with sketches from his 1839 travels. Among these *Intelligencer* articles, one deserves closer scrutiny—an essay summarizing the author's political concerns in 1839. Composed initially as a college lecture, "Things As They Are—Things As We Would Have Them Be" (July 13, 1839) purports to be an exercise in "analytic reasoning," but the display of method is largely a means of justifying the Whig perspective. After enumerating three duties incumbent on every citizen—"moral conduct," "diligent

30. William Darby to D. Christy, February 8, 1839, Manuscript Division, New-York Historical Society, New York City, N.Y.
31. *Ibid.* Darby speaks of correspondence prior to December 1838, in which he advised Christy "to follow Gibbon's with Sismondi's Rome" in his reprint series.
32. *Ibid.*, February 28, 1839.

pursuit of useful knowledge," and an understanding of American government—Darby proceeds to expose a host of evils within and misconceptions about our political system. Renewing an old complaint, he observes that the excessive power of the president has made the government a veritable monarchy rather than a republic. He denounces the extension of suffrage, noting its egregious result: "The vote of the man endowed with the highest attributes of cultivated reason, and adorned with all the graces of unsullied private character, has not one atom more weight than that of the most worthless wretch who staggers to the polls from the intoxicating draught bestowed by a candidate worthy of such suffrage." Shifting his tack, the professor then defends the "American System" of domestic improvement—which had been dismantled by the Democratic party—as a program beneficial to both the landholding classes and the poor. In like manner he praises the National Bank (portrayed by its opponents as "The Monster") as having saved the credit of the country. Interestingly, he criticizes American newspapers for being "vehicles of personal abuse," guided by partisan stupidity—"unbounded and blind confidence on one side, and as unbounded and as blind suspicion and hatred on the other." Sounding at times like a Whig jeremiad, Darby's critique of national life presents a melange of insight and vexation, but above all else it provides a candid reflection of the author's main political concerns at the close of a tempestuous decade.

With discourses on geography, history, and political theory in hand, Darby flew from "flower to flower" during most of 1839. In early March, he lectured at West Alexander (on the National Road between Washington, Pennsylvania, and Wheeling), but he returned to Canonsburg in April for the annual debates of the collegiate societies, Philo and Franklin, and for commencement exercises. In late spring, he made another excursion to Ohio, lecturing on May 27 in Cadiz at the boarding school of Sarah R. Foster and proceeding from there to New Athens, the site of Franklin College. At that academy he lectured in early June, enthralled with the idea that the college stood on land that had been wilderness at the time of his youth. In an *Intelligencer* report, he reflected on the enduring theme of his career, delight with the westward advance of civilization: "When I look back over the last fifty-seven years, I

can scarce credit the evidence of my own senses, or admit the testimony of my own experience. Here am I, sitting in a spacious
room of one of the edifices of a respectable classical institution,
within a few miles of where I have roamed in an unbroken forest.
Now from the window of my writing room appears all the lux-
uriance of rural life in its richest garb."[33] Darby evidently expected
to receive a regular appointment at Franklin College in 1839, but
the post never materialized, perhaps because the school's aboli-
tionist president, Dr. John Walker, learned of the geographer's
"conservative" position on slavery.

As a consequence of this setback, Darby returned to Canonsburg in July, where he composed Tacitus essays at a furious
pace, began papers on "the Great West," and made arrangements
for a grand lecture tour through Ohio. In August, he and his wife
departed, by wagon and riverboat, across the hills and valleys of
Ohio. They reached their first destination, Zanesville, in early
September, and Darby promptly reported to the *Intelligencer*:
"My eccentric peregrinations have brought me to this remarkable spot."[34] What made the area "remarkable" were the geological formations, which he dutifully described; in subsequent
weeks he filed communications from Columbus, Hamilton, Day-
ton, Springfield, Palmyra, and Portsmouth, noting at each stop
the distinctive topographical features and the level of civilization
which these bucolic towns and villages had attained. After three
months on the lecture circuit, Darby reached Chillicothe in early
December, 1839, following an unsuccessful effort to travel up-
stream toward Wheeling (en route to Canonsburg) on the icy
Ohio River. Favorably impressed with Chillicothe, he decided to
spend the winter there, advising his readers in the *Intelligencer*:
"Here is another of those small but splendid centres of rising civilization and prosperity; another Anglo-Saxon triumph over the
wilderness of the West."[35] The town epitomized for him the mi-
raculous influence of learning, culture, and industry on the rude
conditions of western life; struck by the rapidity of change there,
he mused rhetorically at the close of one column (January 31,

33. *National Intelligencer*, June 21, 1839.
34. *Ibid.*, September 10, 1839.
35. *Ibid.*, December 19, 1839.

1840): "Where is now the savage frontier?" The question expresses the essential discovery of Darby's Ohio tour: the same desolate wilderness which he traversed as a young man "under the wing" of Henry Jolly in 1795 had become to his astonishment a thriving, cultivated countryside, everywhere reflecting signs of human progress. On February 20, 1840, Darby witnessed a sight that would gladden the heart of any Whig: "We had a cheering exhibition this morning—nine boats freighted with delegates to the Harrison Convention at Columbus."[36] The spectacle of delegates moving by canal boats to voice support for William Henry Harrison and for internal improvements seemed to Darby a salubrious example of the "energy of Western population." He had seen a sign of the times; the coming national elections would sweep the Democrats out of the White House and give the Whigs a brief hold on the course of public events. A few days after the "cheering exhibition" at Chillicothe, the geographer began the return journey to Canonsburg, where he lectured at Jefferson College during the summer term of 1840. A new decade had begun and, with the success of "Tippecanoe and Tyler, too," the final chapter of his career would soon begin to unfold.

The Jacksonian era had in fact marked the height of Darby's influence and yielded some of his most interesting writing. Viewed as a whole, his work from 1826 through 1840 demonstrates his commitment to three concerns: learning, Whig politics, and the West. Through his books, newspaper columns, and lectures, Darby contributed his share to geographical and historical knowledge, even as he strove to broaden his own education through reading and study. His political instincts aroused by the controversies of the period, Darby turned against his former commander when it became clear that President Jackson had rejected the "American System" of internal improvements, protective tariffs, and a sound banking system. Darby embraced the philosophy of the Whigs and lent his pen to partisan journalism largely because he believed that Whig policies would facilitate the great transformation of the frontier into the "cultivated garden" of America. But his was more than an ideological commitment; as

36. *Ibid.*, March 6, 1840.

his 1836 return to the old border country indicates, Darby never lost the intuitive sense that the meaning of his life's work was bound up with his frontier childhood and with the unfolding phenomenon of the westward movement. In his later years he would grow increasingly nostalgic about the "days of danger and blood" which made possible the settlement of the West, and he would give greater emphasis in his letters, lectures, and newspaper essays to his own role (however modest) in that pioneer movement.

# 5. A Scrivener in the Land Office

For three years, 1838–1840, Darby had immersed himself in his activities as a college professor, lecturer, and traveling correspondent. During this period his daughter Elizabeth married a farmer. named Francis Raikes and in 1840 bore a son named for the father. That same year Darby also learned of a legal controversy in Kentucky involving the estate of his late brother Patrick (who had died in 1829); whether prompted by family concern or cupidity, he resigned his position at Jefferson College in November, 1840, to look into the matter. Toward the end of the year, he departed for Kentucky with Mrs. Darby; while investigating the legal tangles there, he raised money to meet his travel expenses by offering a series of lectures on the "History and Statistics of the United States." If Darby hoped to find himself the beneficiary of his brother's considerable estate, however, he was again doomed to disappointment.

From Kentucky, Darby proceeded to Chillicothe, where (as we shall learn shortly) he spent several aggravating weeks trying to counter the charges leveled against him in the public prints and in legislative assemblies by an expatriate Polish nobleman incensed at Darby's essays on Poland, Russia, and Prussia. During this sojourn, the geographer made an important decision: to leave the West and return to the nation's capital. The Whigs had claimed the White House in 1840, and under the administration of John Tyler (successor to the fated William Henry Harrison), Darby hoped to find some governmental position befitting his age and experience. He had, after all, been a spokesman for internal improvements and the Whig cause for many years, and in his declining years he began to weary of uncertain employment and incessant travel. Apprehensive of the course of national events, he also longed to be closer to the seat of government; increasing sectional tensions seemed to threaten the unity of the country, and Darby hoped through his newspaper columns to advocate reason and cooperation in the face of crisis. This vital work could best be carried out, he believed, in Washington.

By the time Darby reached the capital city, the hopes of the Whig faithful had already been dashed by the unexpected behavior of President Tyler, whose shocking veto of bills to create a new National Bank revealed his states' rights sympathies, splintered the party, and produced a mass resignation of the Harrison-appointed cabinet. The editors of the *Intelligencer*, Gales and Seaton, decried Tyler's actions and began to suspect him of "Locofocoism" (Democratic radicalism). They welcomed the return of Darby, a long-time correspondent and political ally, who shared their belief in sound currency and protective tariffs. The geographer viewed the National Bank setback as a double misfortune: Tyler had repudiated one of the great Whig objectives by use of the presidential veto. Since the early days of the Jackson administration, the veto power had been one of Darby's *bêtes noires*; now the vetoes of a Whig president reinforced those opinions and prompted more vigorous opposition.

Other concerns also occupied Darby's attention in the weeks following his return to Washington. To provide some immediate income, he lectured on August 24, 1841, in the Apollo Hall on the "Anglo-Saxon Population of North America."[1] Armed with statistics from the 1840 census, he endeavored to show that the rapid increase in the white population, particularly along the old frontier, rendered inevitable the settlement of the entire continent within a few decades. While the racial aspect of his theory may trouble the modern conscience, Darby's studies of European his-

1. *National Intelligencer*, August 24, 1841.

tory convinced him that the pinnacle of Anglo-Saxon culture would be reached in North America. Accordingly, he regarded demographic figures as a confirmation of historical destiny, and he wrote often about population patterns in the 1840s, as the nation acquired a new sense of its geographic and spiritual dimensions.

As a measure of their respect for Darby's reputation, Gales and Seaton engaged him to prepare a third edition of his international gazetteer in the fall of 1841. On November 3, they purchased the copyright for $1,200, and on November 29, the *Intelligencer* announced the future publication of "an enlarged and improved edition of the Geographical and Historical Dictionary by Wm. Darby."[2] The notice, almost certainly written by Darby, observed that "there is no man in the country who possesses the same ability, with the historical acquirements and statistical knowledge required for such an undertaking." It also contained a summary of the geographer's ideas on westward movement in America, concluding with a heady prediction: "On this theatre is now advancing the most stupendous of all modern revolutions. An energetic moral mass, augmenting one-third every decennial period, is rapidly spreading westward, and in a few years will present one front to Europe, Western Asia, and Western Africa; and another to Eastern Asia, Polynesia, and Australia." Though the phrase "manifest destiny" would not achieve currency until 1846, five years earlier Darby articulated a similar concept, based not upon a revelation of divine will but upon statistics from the United States census.

Projecting this vision of an Anglo-Saxon nation stretching from coast to coast, Darby hoped his contemporaries would rise above sectional feuds and redefine national interests in terms of cooperation and trust. But the issue of slavery and its extension to new states had already created deep divisions; the middle ground of national purpose seemed to erode beneath Darby's feet. The proposed annexation of Texas disclosed the extent of the polarization: southerners, led by John C. Calhoun, pressed for the admission of Texas to the Union as a slave state, while northern abolitionists

2. Joseph Gales, Jr., recorded the copyright purchase, as well as other payments to Darby, in a ledger now in the Manuscript Collection, Perkins Library, Duke University, Durham, N.C.

argued the immorality of that scheme and questioned the need for westward expansion. In this controversy, the abolitionists were joined by old-line Whigs like Gales and Seaton, who feared that the accession of new territories would cause a sudden transfer of population and capital, producing economic panic. Darby found himself in a curious position during the Texas debate; his experience as a cotton planter enabled him to understand the economic justification of slavery, but he was alarmed by the defiant tone of southern arguments. The influx of American settlers seemed to guarantee the eventual annexation of Texas, but he worried about a depopulation of the Atlantic coast if the government became aggressively expansionist. Darby regarded westward expansion as a relentless, irreversible process, but he wished that his countrymen would view the phenomenon as a national rather than a sectional issue. In a letter to the *Intelligencer* (December 8, 1841) dealing with cotton, he encouraged this larger vision: "There are three changes in progress in North America, over which individuals have no power: 1st, the quadrupling of the Anglo-Saxon nation in each half-century; 2d, the spread of that nation westward; and, 3d, the reciprocal dependence on each other of the political masses of this rapidly increasing People." In a quiet way, Darby attempted to do what many of his day tried (and failed) to do: reconcile North and South by appeals to national pride and purpose.

While he followed the political developments of 1842, Darby continued to prepare his *Dictionary*, sifting laboriously through his sources for the most accurate information. He also did some lecturing early in the year, though the series was discontinued, perhaps due to an illness. During the summer months, Tyler's veto of several tariff bills created a furor in Congress and in the Whig press; Darby, who a year later wrote to a friend "the *Tariff* is all I care about,"[3] joined in the protest by preparing a nineteen-page pamphlet titled *Remarks on the Tendency of the Constitution of the United States to Give Legislative Control to the President*. Carefully avoiding a partisan tone—or direct criticism of Tyler—he once more questioned whether the presidential veto did not in effect destroy the balance of power it supposedly protected. Darby also

3. William Darby to Ephraim Cutler, October 23, 1843, in Peter Force Papers, Manuscript Division, Library of Congress, Washington, D.C.

pointed out the demoralizing result of the congressional vote to override a veto: "Congress is called upon to go through the galling mockery of reconsideration, or, in plain terms, to register sentence of condemnation on its own work, and take upon itself the responsibility, thus relieving the real power. If a more humiliating situation than Congress reconsidering a veto can be conceived of, I must confess it exceeds my comprehension."[4] Darby endeavored to persuade his readers that "the veto in our Constitution is the worst form that power has ever assumed," for "he who can control and arrest legislation must, from the very necessities of such power, have the initiative." He warned that "if the Constitution is not changed, it will change the nation; or, more correctly, it will consummate a change already in progress."[5] In late August, he completed work on his *Remarks*, which evidently appeared a few weeks later.

On August 22, Darby interrupted his labors long enough to write a melancholy reminiscence for the *Intelligencer* (August 27, 1842), occasioned by the death of Henry Jolly, the old frontier scout in whose home Darby had lived in Wheeling in 1794. Written out of "duty . . . to the last remnants of our heroic age," the essay teems with recollections (many concerning the author) stretching back half a century. In contrast to the febrile 1840s, the earlier period indeed seemed "heroic" to Darby, who believed the hunter-warrior possessed a type of courage that had vanished from the American scene. Though not the greatest of these heroes, Henry Jolly was better educated than most and possessed "ideas and habits of caution and discipline" that "early pioneer warriors" generally lacked. In the coming years, Darby would look back often to frontier days in the Ohio Valley with the poignant realization that he was among the last who could recall the settlement of "the Western country."

Probably to improve his financial situation, Darby resumed his lectures on "Astronomy, Geography, and History" on September 8, 1842. These addresses, delivered at his "Lecture Room" on New York Avenue, continued through the fall months, with a

4. *Remarks on the Tendency of the Constitution of the United States to Give Legislative Control to the President* (n.p., 1842), 4.
5. *Ibid.*, 19.

subscription for the entire series priced at five dollars.[6] Soon after the lectures began, Darby obtained a humble government position: on September 21, he began daily work as a clerk in the General Land Office at an annual salary of one thousand dollars.[7] There is a pathetic irony in the appointment; one of the nation's leading topographers, an authority on the trans-Allegheny West, had been hired at a miserable salary to transcribe documents pertaining to land surveys and sales. But when the *Monthly Journal of Agriculture* portrayed him in August, 1845, as a victim of injustice, "employed this hot weather, through his 'ten hours,' at one of the lowest desks in the Treasury Department building, on the pay of a half-fledged midshipman, at work that any common clerk might perform," Darby politely corrected the inaccuracies (his office was on the same floor as that of the Land Office commissioner) and insisted: "With my situation I am perfectly well satisfied."[8] Characterized by the *Intelligencer* (June 21, 1843) as an "unostentatious" and "unassuming" man, Darby seems to have accepted his menial duties with gratitude; he retained the position—at least nominally—for the last twelve years of his life. A short biographical sketch from the New Orleans *Commercial Bulletin* (June 12, 1843) portrays the government clerk, at the age of sixty-seven, as his contemporaries knew him:

> Mr. Darby is a man of enlarged mind, cultivated, adorned, and furnished through many years of toil. His life is a model of patient and persevering industry. Few men have deserved as much from his fellows, and few men have received as little.—He is a true patriot, as is attested by his conduct in war, and by the spirit which breathes through all his writings. He is a philosopher, a political economist, and historian, geographer, astronomer, and mathematician of the first rank; but he is simple, plain, and unostentatious in his habits— studious, retiring, and modest—and, accordingly, is less known than thousands with not a tithe of his pretensions.[9]

6. *National Intelligencer*, September 2, 1842.
7. U.S. Cong., *House Documents*, 27th Cong., 3rd sess. (Washington, 1843), V. Doc. 160, p. 28.
8. *Monthly Journal of Agriculture*, I (August, 1845), 109. Darby's correction appeared in the Washington, Pa., *Reporter*, January 3, 1846.
9. The profile was reprinted in *Niles' Weekly Register*, LXIV (July 1, 1843), 277.

A letter written by Darby and published in the *Daily Madisonian* (January 10, 1843) affords an illuminating example of his capacity for principled judgment. The communication treats a minor but controversial issue then current: the campaign to remit to the ailing, debt-ridden ex-president, Andrew Jackson, the fine imposed upon him in New Orleans in 1815 by Judge Dominick Hall (see Chapter 2). Congressman Henry A. Wise of Virginia had written to Darby on January 3, 1843, requesting an "impartial history" of the events surrounding the declaration of martial law, noting that the geographer had been in New Orleans "at the time of the defence" but had never been "a sycophant" of the general. In his reply, Darby took pains to declare: "General Andrew Jackson never conferred on me the slightest favor." The remark savors of the resentment which arose between 1825 (when Darby appealed unsuccessfully for the hero's assistance) and the rancorous newspaper exchanges of 1835–1836. Yet Darby felt compelled to defend Jackson's high-handed actions in emphatic terms: "He took the straight path and followed it, and if he jostled those who stood in his way rather violently, he persevered and saved his country." Complaining of the "injustice" and "ingratitude" of Judge Hall's sanction, Darby contrasted the gallantry of Jackson with the judge's craven retreat from the city during the fighting:

> I have lived to see no slight attempts made to cast odium on the head of this veteran, to lessen the value of his services, discolor his motives, and to insinuate that those who with him braved the danger, and achieved a victory which shed beams of glory over the nation, were supple time-servers, and that the great Hero of that day was Dominic A. Hall, who did not even leave the light of his countenance to cheer the real heroes.

There is in the defense of Jackson a subtle note of self-glorification: Darby was among "the people of Louisiana" who fought "to defend their homes, their daughters, wives, and their sisters—all that was most dear—from insult, from violation: to chastise an insolent and too often ruthless enemy, and to close a war in a halo of glory, which, had Louisiana submitted to conquest, must have terminated in a gloom of shame from the Passamaquoddy to the

Sabine." This expression of pride nevertheless serves to underscore a point crucial to the argument for martial law—the British forces posed a powerful threat to Louisiana (the Treaty of Ghent notwithstanding) and their continued operations along the Gulf Coast after the great battle of January 8 justified the civil vigilance that Jackson sought to enforce. Darby's was one of many voices raised in the general's behalf during this belated campaign for justice; on February 14, 1844, President Tyler signed a bill approving full remission of Jackson's fine, with interest, and thus removed an onus which had long hung over the general's famous victory.

As for Darby, his reputation and influence faded in the early forties, though he continued to defend his published opinions with a fierceness that belies the image of a "retiring" scholar. No episode illustrates this facet of his personality better than his lengthy feud with Major Gaspard Tochman of Louisville, Kentucky. A Polish immigrant of aristocratic birth, Tochman had written to the *Intelligencer* in October, 1839, attacking the opinions of Tacitus and suggesting that the author of the "Northern Nations of Europe" series must be a member of the Russian Legation or—worse— a "secret agent" of the Russian government. Tochman refuted Darby's assertions that the mass of Poles were oppressed by the ruling classes of Poland and that Russia enjoyed greater prosperity because of its more enlightened political system; he had, moreover, prepared a series of articles exposing the deceitful, nationalistic motives of Tacitus, but Gales and Seaton refused to publish them, as they were founded on a radical misconception of the correspondent's purposes. The Pole then took a more aggressive approach, going on a lecture tour through Kentucky and Ohio to excoriate Tacitus and whip up sympathy for the misunderstood Polish nobility. Tochman even spoke (according to Darby's report) in legislative halls, where "members of Legislatures" reputedly joined with "other prominent citizens" to form preambles and pass resolutions of censure against Darby.[10]

Such response to his ponderous newspaper essays probably astounded Darby, who seems to have encountered public reaction for the first time in late 1840. While in Kentucky attending to his

10. See Darby's advertisement for *The Northern Nations of Europe* in the *National Intelligencer*, March 20, 1841.

brother's estate, he evidently met with open hostility during his lectures, and he shortly received news of yet another campaign against him, ironically launched by Tochman in that center of "rising civilization," Chillicothe. There, the Pole had railed against Tacitus, boasting that he had "silenced that bell" on the subject of Poland. Never one to suffer an insult passively, Darby departed at once for Chillicothe to rescue his reputation. By the time he arrived, Tochman had disappeared, and so the geographer countered his defamatory speeches by publishing, in March, 1841, a collection of *Intelligencer* essays entitled *The Northern Nations of Europe, Russia and Poland*, offered (as the author put it) in "self-defense." Clarifying his motives, Darby noted in the preface: "When the communications signed Tacitus were written by me in 1839, they were prepared to show my countrymen, the opposite consequences of order and disorder, by means of most striking examples afforded by modern history, the causes of the Rise and Progress of Russia, and Decline and Fall of Poland." Obviously stung by the support that Tochman had generated, he warned the public: "To you, I may observe, while laying the following sheets before you, that in sympathizing with the Polish nobility, you choose the oppressors in place of the oppressed."[11] The book, issued in Chillicothe and priced at a mere twenty-five cents, appeared in early April and seemed to vanquish his adversary; however, Darby had not heard the last from Major Tochman.

Two years later, Darby chanced to see in Washington a copy of the Richmond *Compiler*, which reported the substance of an address by Tochman before the House of Delegates. Darby could not refrain from public comment; in a Tacitus essay titled "Poland" (March 31, 1843), he disputed Tochman's claim that "there were no counts or titled nobility" in Poland at the time of its decline, and he cited sources to confirm the point. The article touched off another skirmish with the exiled Pole, for on April 20, the *Intelligencer* printed a response from Tochman which repudiated Darby's authorities and attacked the geographer personally. The newspaper omitted portions of the letter impugning Darby's

---

11. "To the People of the United States," *The Northern Nations of Europe, Russia, and Poland* (Chillicothe, Ohio, 1841), iii–iv.

patriotism, even though the rival *Daily Globe* had printed the entire rejoinder a day earlier.

Determined to silence his nemesis, Darby replied with an open letter to Tochman (April 25, 1843), chiding him for "personal abuse" and issuing a proposal: "I openly challenge you, Major Tochman, to come to this city as soon as your convenience will admit after the opening of the next Congress; we can then appear before persons from every section of the United States. If you accept such alternatives, I am ready to meet you, in open assembly, on the following terms." Carefully avoiding the suggestion of a debate, Darby stipulated that each man would speak for one hour, with Tochman deciding the order of appearance. He then shifted to the irksome matter of Tochman's supporters and questioned their historical understanding, warning that if "the designs of such men as Major Tochman" were adopted, the certain consequence for the Polish people would be "blood, fire, and utter ruin." Announcing that he would not be "bullied into a newspaper controversy," Darby called on his adversary to "decide matters more honorably than by abuse." Tochman accepted Darby's challenge to meet in "open assembly"—which he construed as a debate—asking only that the scope of discussion be broadened. This request Darby quickly refused (*Intelligencer*, May 24, 1843), insisting that the disputants argue the decline and fall of Poland in the specified manner. The refusal went unanswered, but Tochman had not been muzzled.

Meanwhile, Darby neared completion of his gazetteer. On June 10, the *Intelligencer* announced that the volume had gone to press, "having undergone the most extensive revision by the eminent and skilful [*sic*] Geographer whose name it bears." Gales and Seaton attempted to puff the work by publishing lengthy excerpts, such as the article on Ireland (July 13, 1843), which Darby unhesitatingly termed "the finest island on the earth." By mid-October, over seven hundred pages were in print, leading the author to write a friend: "The Day is breaking on which our long promised Dictionary is to appear, and glad will we all be at the consummation." [12] Believing that a Whig defeat in the next election

12. William Darby to Ephraim Cutler, October 23, 1843, in Peter Force Papers, Manuscript Division, Library of Congress, Washington, D.C.

would eliminate his government position, he considered resigning in 1843 to promote the sale of his book in the West but decided at length to "await the issue and act upon events." [13]

Both the Whigs and Democrats were torn by deep divisions in 1843, and Tyler seemed a hopeless case, having lost the support of his own party. Viewing the Washington turmoil, Darby sensed that a "day of revolution" lay ahead and observed: "How far the revolution may affect me is of course beyond our ken, but the revolution itself is sure. I do not hesitate to say that Mr. Tyler has been rather ungratefully treated by both parties. His Veto of the Bank saved the Democratic party, and his signing of the Tarif [sic] saved the credit of the country. But his doom is sealed unless a great and unlikely change is effected. Again, unless there is more unanimity than has recently been manifested by the Democratic party, its fate is sure." [14] In the same letter, Darby paid tribute to "the much abused Alexander Hamilton" and affirmed his concern for the tariff, noting that "public prosperity has risen and fallen, just as this primary policy has been followed or abandoned." It mattered little to him who the new president might be, so long as the "primary matter" of the tariff was secure.

Amid the political uncertainty of the season, Darby was solaced by a visit from his daughter and son-in-law, Elizabeth and Francis Raikes, who had come from Ohio with their three children (a grandson and namesake of the geographer had been born in 1842) to await the birth of their fourth. Elizabeth Darby seemed "in better health than usual" during her daughter's visit, despite an attack from "the Fall fevers" which had gone the rounds in the small house on Seventeenth Street. The atmosphere of cheerful anticipation—Darby awaited a grandchild and a new book simultaneously—was broken, however, when he glanced at his copy of the *Intelligencer* on October 21. There he found a notice of two lectures to be given in Georgetown by Major Gaspard Tochman on "Poland and Russia"; without warning, his long-time adversary had arrived in the city, perhaps seeking the public confrontation that seemed an inevitable consequence of their battles in print.

On the evening of October 23, Darby attended Tochman's lec-

---

13. *Ibid.*
14. *Ibid.*

ture in Georgetown, presumably to measure his opponent. There the two met for the first time, and a few days later Tochman visited Darby's home, subsequently remarking: "The kind and hospitable welcome and reception which he and his family gave me have increased my esteem for my learned antagonist."[15] Yet Tochman had come to fight; in defense of the Polish nobility, he offered a series of lectures to win over the public prior to his "debate" with Darby. His use of the word in an advertisement (*Intelligencer*, October 28, 1843) sparked another flurry of semantic and scholarly wrangling, capped by Darby's charge (*Intelligencer*, November 9, 1843) that Tochman was either "ignorant of the history of his country" or that he had "willfully misrepresented it." But despite the warning, Washington lionized the courtly foreigner; even the editors of the *Intelligencer* spoke of "an interesting and eloquent address" by Tochman (November 13, 1843). However, when he once more assailed Darby's scholarship (*Intelligencer*, November 14, 1843), Gales and Seaton announced that, after the geographer's reply, the controversy would be "considered as closed," since the newspaper had already given it attention "disproportionate to its immediate interest or consequence." The barrage of facts, references, and retorts had evidently numbed everyone except the two participants. With Darby's somewhat pitiful final letter (*Intelligencer*, November 21, 1843), in which he pleaded for sympathy on the basis of his Irish ancestry, the feud came to an end. Though he apparently lost the popularity contest, Darby at least demonstrated his tenacity and his passion for historical truth. He also displayed some pique: despite his professed admiration for the Polish people, he never understood how his compatriots could side with an exiled Pole.

The quarrel with Tochman came to an end just as the *Intelligencer* hailed the appearance of Darby's new *Universal Geographical Dictionary* (November 29, 1843). Predictably, Gales and Seaton's reviewer lavished praise on the author, listing his principal qualifications as "a laborious life consecrated to study and travel," "a wide range of information and research," "a high competency in such physical sciences as can aid geography," "a sincere love of truth,"

15. *National Intelligencer*, November 6, 1843.

and "a heart genuinely American." Had he written the notice him-self (which seems unlikely), Darby could not have devised a more generous encomium. The *Dictionary* proved to be his last major publication, and though its commercial success was not great, the gazetteer illustrated, in its announced emphasis on "the Central States and Territories of the United States," the same fascination with the West which informed Darby's earliest works.[16] But while westward movement had once implied simply a civilizing of the backwoods through internal improvements, it now represented in his view the final phase of a vast historical process, "the most stu-pendous of all modern revolutions."

Darby followed this revolution with keen interest, for he saw the opening of the Far West as a consequence of astonishing devel-opments in transportation and communication. Citing one inven-tion vital to western growth, he wrote: "I have seen this steam changing the history of the world, and placing the name of Fulton in the list which can perish only with time." He continued, "We have now before us the name of Morse, who has given the motion of light to thought."[17] But the development which seemed most crucial to the settlement of the West was of course the railroad. Darby became an early advocate of a railroad to the Pacific coast, forecasting in the *American Review* (April, 1845): "Though we cannot hope to enjoy such a jaunt, many, we make no doubt, are the children now in life, who will pass on railroads from the tide-margin of the Atlantic to the tide-margin of the Pacific."[18] The need for such a railroad was amply demonstrated, Darby believed, by population trends: "The Anglo-Saxon population of North America is rapidly advancing toward an entire, irreversible change in the history of the world." He concluded that "by wise foresight and true statesmanlike principles of action, trusting less to force than to the steady and inevitable triumphs of Time," the western boundary of the nation would safely become "the Pacific shore."[19]

16. The book's poor sales are alluded to in a brief notice in the *National Intelligencer*, December 11, 1845: "This Treatise, through the third edition, and nearly two years in the market, remains but partially published, and its contents in great part unknown."
17. *National Intelligencer*, October 13, 1846.
18. *American Review* (April, 1845), 432.
19. *Ibid.*

Albeit indirectly, Darby asserted that a Pacific railroad would ensure the migration of Americans to the West, enabling the country to establish territorial claims through the sheer force of numbers rather than by bloodshed. This was, in effect, a protest against the militaristic attitude, rampant in 1845, that the United States should take Texas, California, and the intervening land from Mexico by any means necessary.

Darby's connection with the *American Review* deserves consideration at this point, since George H. Colton's New York journal commanded immediate attention in 1845 by virtue of its quality and variety. Among the pieces appearing in the first volume (January–June, 1845) were several works by Poe including newly revised texts of "The Valley of Unrest" and "The City in the Sea," two tales by the young Walter Whitman, essays by Melville's mentor, Evert A. Duyckinck, a "Review of the 28th Congress" by Horace Greeley, and a second essay by Darby titled "Prussian Empire." Later in the year Darby contributed lengthy essays on Frederick the Great and Prussia, the latter appearing in the same issue with Poe's "The Facts in the Case of M. Valdemar." Allied with the *Knickerbocker* in partisan opposition to the *Democratic Review*, the *American Review* offered an impressive selection of poetry, fiction, criticism, and political and economic commentary. It drew upon the talents of many writers of the Whig persuasion; even its book reviews were charged with party feeling.[20] Though a regular contributor for only one year, Darby published two subsequent essays in the journal, both concerning that preoccupation of his final years, the Pacific railroad.[21]

In 1845 Darby again became an occasional contributor to the *Intelligencer*, after a year's respite. In an essay on the "Progressive Population of the United States" (January 2, 1845), he worked out an elaborate projection of national growth through the year 1901.

20. Though associated with the "Young America" group (mostly rabid Democrats), Duyckinck and Poe were welcomed in the Whig journal in 1845, since political events had created a temporary alliance between Locofocos and Whigs, who both opposed the annexation of Texas and war with Mexico. See Perry Miller, *The Raven and the Whale* (New York, 1956), 121–26.

21. See "The Great Pacific Railroad," *American Review*, X (September 1849), 311–13, and "Pacific Railroad," *American Whig Review*, XII (November, 1850), 539–46. Both essays review and endorse a plan by Asa Whitney to construct a railroad from Lake Michigan to the Pacific.

Working from census figures from 1790 to 1840, he identified what seemed to be a fixed ratio of population change—an annual increment of 3 percent—and calculated the consequent increase through the end of the century. Compared with the eventual census figures, Darby's projections through the year 1870 proved highly accurate (within 2 percent), but his estimations for subsequent years reveal an increasing inaccuracy. By the year 1900, his projection exceeded the actual population by 30 percent. However, what seems noteworthy is not Darby's statistical performance but his anticipation of a seemingly incredible increase in population with attendant changes in national life. In making such predictions, he risked "the froth of the public current, *ridicule*,"[22] but having witnessed "wonders incomparably more remarkable" in his seventy years, Darby seemed to enjoy the role of prophet— especially since his prophecies sprang from computation rather than conjuration.

His "public writing" for this period reflects three centers of interest: demography, European history, and astronomy. In a letter to the *Intelligencer* (June 12, 1845) he repeated his contention that "the law of progression of our population is founded on simple arithmetic principles." In November he prepared an article comparing the population of different regions of the country for John S. Skinner's *Monthly Journal of Agriculture*. In that essay (published in January, 1846) Darby expressed concern about the "far too prevalent emigration from the Atlantic region . . . into the interior and Western sections of North America." But he acknowledged the attraction ("Land! more land!") and judged that "to stay the current of Western emigration is a hopeless project." To the consternation of Skinner and his eastern readers, Darby concluded, 'If no change takes place in the current of emigration, the centre of political power must correspond with the centre of force, and leave at long distance the Atlantic coast."[23] Thus Darby anticipated the kind of depopulation he had portrayed a decade earlier in the closing lines of "The Wedding." Political power of a different sort formed his subject in three papers on Prussian history (June

22. *National Intelligencer*, June 12, 1845.
23. "Comparative Views of the Progress of Population in Certain Regions of the United States," *Monthly Journal of Agriculture*, I (January, 1846), 376.

10, July 2, 1845; February 20, 1846), which paralleled his essays for the *American Review*. He dealt at length with the reign of Frederick the Great, whom he considered "perhaps the most tolerant man in Europe," a principled ruler who achieved civil order without tyranny. Darby's passion for order seems also to have inspired his third interest, astronomy. The constancy of the stars provided a reminder of eternal truths and a way of uniting past and present imaginatively: "When we behold the constellations, we know that our race, from the most remote ages, gazed on the same stars thus grouped, and we know also that all future generations will contemplate the same objects, in the same relative position; and thus the poetic, historical, and scientific applications of this preeminent science unite all times and all countries, and all nations who have risen or may rise to the rank of civilization."[24] Subsequent articles (February 7, February 19, 1846) demonstrated a lively interest in the new planets which contemporary astronomers believed that they had found.

In late summer, 1845, a separate and significant development rekindled Darby's enthusiasm for frontier history. On August 1, he received through his friend at the General Land Office, Colonel Samuel H. Laughlin, a letter from a young scholar named Lyman C. Draper, containing an ingratiating request for the geographer's remembrances of pioneer days in the Ohio Valley. Excited by the inquiry, Darby wrote immediately to Draper: "You do me no more than justice when you suppose that I take a deep interest in whatever relates to the Great West. In fact all my other thoughts are incidental; the West is the home of my imagination and warmest recollections and aspirations."[25] Draper's desire to collect material for a comprehensive history of the frontier won Darby's instant confidence and admiration. During the next five years he corresponded frequently with the younger man, supplying the names of early settlers, recounting events of historical interest, and urging Draper on with his valuable work. Largely through these letters, the details of Darby's early years have been preserved; they also contain intimate glimpses of his personal life from 1845 to 1850. Though Draper never finished the projected

24. *National Intelligencer*, October 13, 1845.
25. Darby to Draper, August 1, 1845, in Draper Manuscript Collection.

study, his letters and documents from countless correspondents comprise to this day the most valuable single collection of manuscripts pertaining to the early West.[26]

Darby took seriously his commitment to assist Draper, for, as his letters suggest, he considered himself one of the pioneers most deserving of a place in the contemplated work. He wrote to Draper, "We old warriors are delighted to clutch the pen and in fancy return to the battlefield,"[27] blithely ignoring the fact that he had played no part in the border wars of Pennsylvania and Ohio. In early 1846 he established communication with two old Wheeling acquaintants, Mrs. Lydia Cruger and Lewis Bonnett, Jr., and reported their recollections to Draper, assuring him: "All I now can do I will do, to forward and give aid to your noble enterprize."[28] With increasing frequency he pressed the historian for a prompt publication of his work, impatiently asking in one letter, "Your Book—your Book. When will we be able to say we have read the History of the Heroic Age."[29]

In October, 1846, Darby and his wife moved from Washington City to the village of Georgetown, taking a house on Second Street a few doors east of Georgetown College. The government job no longer required his daily presence; as he explained to Draper: "I do my public writing at home and visit the Land Office only when it is necessary to my business."[30] As it happened, Darby spent much of the spring and summer of 1847 away from his desk in Washington, due to the "severe and protracted illness" of Elizabeth. Depressed by his wife's decline, he found even his infrequent visits to the Treasury Department tedious; only his correspondence with Draper offered a "restorative" after "the dull round of office labor."[31] Elizabeth's condition grew worse, and in

26. Volume VIII of the Pittsburgh and Northwest Virginia Papers in the Draper Manuscript Collection includes twenty letters from Darby to Draper: two written in 1845, seven in 1846, six in 1847, one in 1848, and four in 1850. It also contains one letter from Darby's third wife, Mary, to Draper. See my "Glimpses of the 'Heroic Age': William Darby's Letters to Lyman C. Draper," *Western Pennsylvania Historical Magazine*, LXIII (January, 1980), 37–48.

27. William Darby to Lyman C. Draper, August 19, 1845, in Draper Manuscript Collection.

28. *Ibid.*, July 24, 1846.

29. *Ibid.*, January 9, 1847.

30. *Ibid.*, January 18, 1847.

31. *Ibid.*, May 21, 1847.

mid-summer Darby wrote to his daughter in Ohio, urging her to hasten to her mother's bedside. But before Mrs. Raikes could reach the capital, Elizabeth died on July 31, 1847, at the age of sixty-two. For thirty-one years, she had been Darby's "most affectionate companion," accompanying him in nearly all of his travels through Pennsylvania, Ohio, and Kentucky; her death grieved him deeply. Following the funeral, Darby accompanied his daughter to Ohio, where he remained for about two weeks. After a brief, nostalgic visit to Washington, Pa., he returned to Georgetown in September to put his life in order again.

Through his public duties, his friendship with Colonel Laughlin, and his correspondence with Draper, Darby endeavored to regain his equilibrium. Acknowledging the "utter derangement" of his "domestic situation," he revealed his confusion and sadness to the historian: "In truth, my Dear Friend, my mind wants recovery from the shock. Complete recovery is beyond hope. Amid the trial I entered my 73rd year, a time of life when Hills in our way become mountains. It is true and most consoling is the Truth, that my friends have most feelingly soothed the pain of such a loss— But the charm of my life is broken, gone."[32] Coupled with his emotional loss, Darby apparently found himself in financial difficulty toward the end of 1847. His predicament compelled him to take an action he had long resisted: an appeal to Congress for financial relief.

At his own expense Darby published a statement titled "Notes in Regard to my Survey of the Sabine River" in December, 1847, describing the topographical investigation which in effect fixed the boundary between Texas and Louisiana.[33] He also explained how John Melish, the Philadelphia map maker, had deprived him of the profit and acclaim to which he was entitled. In mid-January, 1848, Darby submitted his petition to Congress, avowing that "ten thousand dollars would not be an over charge" for services rendered to the government but agreeing to accept gratefully "whatever amount may be awarded." A House committee swiftly rejected the claim, noting that whatever injury Darby had received from Melish, the government could not be held responsi-

32. *Ibid.*, September 7, 1847.
33. See O'Rielly, "Pioneer Geographic Researches," 223–25.

ble. The Senate Committee on Claims, however, found that the government had "availed itself of the unpaid labors of Mr. Darby" and reported a bill on August 9, 1848, allowing a $1,500 compensation.[34] For some reason the bill never reached the Senate floor, but Darby persisted and a new petition for his relief was introduced on January 14, 1850, by Senator Solomon Downs of Louisiana. The entire Senate debated Darby's case on April 23, 1850, at which time Downs praised the claimant: "Mr. Darby has rendered important services to the country. He has done more perhaps to develop the geography and resources of the West and Southwest than any other man. He has added much also to American literature, in the preparation of his geographical and statistical works. . . . He now finds himself, in his extreme old age, very poor, occupying the third story of some Government building, in the discharge of the duties of some petty office, for which he receives the little pittance of $1000 a year." After considerable discussion of the government's obligation to Darby, the bill passed, and on May 9, it was referred to the House, where it died in committee.[35] Again in 1852 the Senate sent a bill on to the House, but it disappeared in committee once more. Finally the Senate Committee on Claims reported a bill on April 19, 1854, authorizing compensation for the geographer, and on July 7, the Senate passed the measure without amendment. The House this time followed suit, and on August 1, 1854, President Franklin Pierce signed the joint resolution, an "Act for the Relief of William Darby."[36] Ten weeks before his death Darby received a $1,500 recompense for his toils in Louisiana forty years earlier.

His determination in pursuing the appeal is characteristic of the vigor which marked his later years. Though the death of Elizabeth stunned Darby and temporarily disordered his life, by spring, 1848, he was again contributing essays on European history to the *Intelligencer*. As he admitted to Draper, his heart was not really in

34. U.S. Cong., Senate, *Reports of Committees*, 30th Cong., 1st sess. (Washington, 1846), I, Rept. 236, p. 3; U.S. Cong., House, *Reports of Committees*, 30th Cong., 1st sess. (Washington, 1847), I, Rept. 154.
35. John C. Rives, *Congressional Globe*, 31st Cong., 1st sess. (Washington, 1850), XXIII, Part I, 150, 805, 973.
36. John C. Rives, *Congressional Globe*, 32nd Cong., 1st sess. (Washington, 1852), XXIV, Part I, 199, 207–209, 224, 289, and 33rd Cong., 1st sess. (Washington, 1854), XXVIII, Part II, 943; Part III, 1628, 2039.

the work: "Except what the public prints present you with dayly, my life goes on now so much in routine that I have nothing of moment to fill a page." The routine was broken, ironically, by another death; when Darby journeyed to Baltimore to console the family of his brother-in-law, Benjamin Tanner (who died November 14, 1848), he renewed his acquaintance with Tanner's daughter Mary and formed a quiet affection for her. On April 10, 1849, the two were married in Baltimore by the Reverend Thomas Atkinson. In a letter to Draper, Darby recounted his marital history and tactfully compared his "excellent" third wife to her predecessors: "The present cannot be superior but is equal to either."[37]

His marriage to the niece of his second wife had a tonic effect on Darby's health and productivity. During 1849, he composed a dozen essays and letters for the *Intelligencer* and an article for the *American Review* (September, 1849). His newspaper contributions aptly illustrate the direction of his thinking during his later years—away from immediate, controversial issues and toward the universal principles which govern natural phenomena and human history. His disinterest in partisan affairs manifests itself in a letter (*Intelligencer*, June 22, 1849) defining his political stance: "In regard to the primary measures of the Whig party I assent fully, and, so far, have been, am now and trust will remain for life a Whig; but, into party politics, have never, and while reason serves, never will enter." Darby apparently referred to office-seeking, for he had been active in partisan journalism. In "The Past and the Future" (*Intelligencer*, November 1, 1849), Darby magisterially examined the "Rise, Progress, and Fall of Empires," noting "a steady obedience to an irrepealable law" in the unfolding of history: "What are the components which form the mass of element for history? Why, the recital of struggles which wear our strength away; attempts of the strong to oppress the weak. Follow the analysis, and we find the strong losing strength, and the weak becoming strong. Is either taught by their own experience? . . . Echo is silent. No response to such questions."

Darby entered his seventy-fifth year in possession of sound health and an unimpaired mind. In 1850 he and his wife Mary

37. William Darby to Lyman C. Draper, April 10, 1848, and April 10, 1850, in Draper Manuscript Collection.

moved back to Washington City, taking a house on New Jersey Avenue near the Capitol. All of Washington was "in agitation" over the famous compromise proposed by Henry Clay to settle the issue of slavery in the West; without specifying his objections, Darby wrote to Draper: "From this Compromise, I hope so little that I cannot well be disappointed."[38] The death of Colonel Laughlin, on May 6, 1850, deprived Darby of his closest friend, but he assuaged his sorrow through hard work, producing some lengthy essays for the *Intelligencer* and developing plans for a new, revised edition of his gazetteer. On September 18, 1850, he wrote to the Philadelphia publisher Henry Carey, proposing "either a Gazetteer of the U.S.; or a general Geography of the World." Darby planned to incorporate the results of the U.S. Census of 1850 and assured Carey that he was "much better situated, to undertake and execute the labor of such compilations" than he had ever been.[39] But the projected volume never materialized, perhaps because Carey knew the commercial fate of the 1843 Gales and Seaton edition.

Darby also renewed his correspondence with Draper in 1850; after a silence of two years, the geographer, in response to a letter from Draper, expressed his "sincere pleasure" in learning that the historian was "preparing the rich legacy" he would "bequeath to coming generations."[40] But his impatience can be seen in a subsequent letter: "When will your work appear? . . . You . . . owe the Debt to the world, and should make every reasonable exertion to fulfill your engagement, lest accident should mar the Sacred Duty. As the Oldest Living Man whose memory mingles with the great Band; in their name I say, haste and place those names on a Tablet durable as their services were deserving."[41] As this passage suggests, Darby had come to see himself, in 1850, as a lone survivor of the "Heroic Age." For several years, one of his self-appointed duties with the *Intelligencer* had been that of memorialist for fallen pioneers. His obituary notices of such persons as Henry Jolly (Au-

---

38. William Darby to Lyman C. Draper, May 9, 1850, in Draper Manuscript Collection.

39. William Darby to Henry C. Carey, September 18, 1850, in Simon Gratz Collection, Manuscript Division, Historical Society of Pennsylvania, Philadelphia, Pa.

40. Darby to Draper, April 10, 1850, in Draper Manuscript Collection.

41. *Ibid.*, December 4, 1850.

gust 27, 1842), Mrs. Ephraim Cutler (July 28, 1846), Martin Du-
ralde (December 5, 1848), and George King (January 22, 1851) afforded an opportunity to reminisce and to remind the public of its
obligations to the "patriot pioneers, who, in the face of privation,
danger, and death, wrested Central North America from the savage, and changed the wilderness to a garden."[42] In late autumn,
1850, Darby delivered a series of lectures on the pioneers, about
whom he felt peculiarly qualified to speak: "I'll forget the Pioneers
when I forget my own parents, my Brothers and Sisters, and myself, for such we all were."[43]

Through an unusual arrangement, Mary Tanner Darby assumed her husband's duties at the Land Office in 1851, though the
geographer's name continued to appear on the official list of em-
ployees. But if age and infirmity kept him from his government
work, it did not significantly interfere with his studies or his public writing. Darby took a powerful interest, for example, in the
results of the 1850 census, which seemed to corroborate his theory
of population increase. He translated raw statistics into an optimis-
tic vision of the future:

> The spread of Anglo-Saxon population over the great central zone
> of North America, if taken alone, would rank as one, if not the
> most important one, of the permanent changes in the condition
> of our race; but, when combined, on a continent presenting two
> oceanic fronts, with the railroad means of locomotion and tele-
> graphic rapidity of thought, and one people, with a common and
> energetic language, imbued with similar views on political and civil
> government, and also of the principles of moral conduct, an ad-
> vance and permanency of human prosperity and happiness may be
> rationally hoped for, on an extent of surface never before realized.[44]

During his final years, Darby sounded the theme of population increase repeatedly, urging statesmen and citizens to ponder the inevitable changes in national life and domestic policy which it
would bring about.

As he looked to the future, Darby also reflected upon his frontier boyhood in several *Intelligencer* articles of 1851–1852. One es-

---

42. *National Intelligencer*, January 22, 1851.
43. Darby to Draper, December 4, 1850, in Draper Manuscript Collection.
44. *National Intelligencer*, April 17, 1851.

say (August 7, 1851) concerns the changing conception of the phrase "the Great West": seventy years earlier the name had designated the country beyond the Monongahela River, but at mid-century it referred to the territory west of the Mississippi. In a letter ostensibly about the value of internal improvements (November 6, 1851), the geographer recounted the conditions which families like the Darbys faced on the early frontier:

> Few and far between were the habitations of the pioneers. The rude but very necessary stockade fort or blockhouse [stood] every few miles amongst the also rough log-cabins of the settlers. Such was the country into which, as if impelled by unseen power, pioneers pressed, braving the most fearful enemies and the labor of conquering that enemy and a dense forest. The savage foe and the forest fell before them. Honor and veneration is their due; they were the advance guard of civilization; and many were the parents of the living generation.

If he considered the white man's subjugation of his "savage foe" an inevitable happening, he now felt some compassion for the Indian, advising in one essay (September 14, 1852): "Let the trembling remnants of once savage tribes be dealt with tenderly but wisely." Specifically, he advocated the assimilation of tribes into the general population, fearing that "left in groups," the Indian's "gloomy energy" would give way to a "stupid, idle vacancy" and hence collective paralysis. In making such pronouncements, largely for the benefit of a younger generation unfamiliar with frontier life, Darby often represented himself as one who had witnessed the entire course of his country's history.

Darby made a final journey to "the Western country" of his youth in the summer of 1852, visiting central Ohio and "its really flourishing capital, Columbus."[45] He probably traveled by railroad, along routes which had once defied the covered wagon; instinctively he must have formed mental contrasts between the passing countryside and the wilderness of an earlier day. After partaking of his daughter's hospitality, Darby returned to Washington in early September, where he resumed his writing and made plans for a series of lectures on comparative geography, which he deliv-

45. *Ibid.*, September 14, 1852.

ered that autumn. If he took an active interest in the presidential election of 1852, he never made public his sentiments.

By 1853, Darby's salary in the Land Office had risen to $1,100 (an increase of one hundred dollars after a decade of service), which provided him a modest subsistence. He perhaps received occasional remuneration from the *Intelligencer*, which published at least ten articles and letters from his pen during that year. Several essays reflected a new topic in his writing—the prospect of war in Europe. Pointing to the maxim "long peace, long war," Darby argued that the continent was ripe for conflict and predicted a clash between Russia and her neighbors. In one essay (July 28, 1853) he warned: "The threatenings of war now are not unsubstantial sounds. They are the echoes of a voice which has resounded through all ages. They are the angry threats of human passions. In the mean time tears and blood sink into the earth, and the moanings of humanity are wasted on thin air. All ages afford evidence of these truths." Later (August 25, 1853), he foretold "a most extended, destructive, and, in duration, perhaps unexampled war" in Europe. Interestingly, he seems also to have sensed an inevitable confrontation between Russia and the United States; pointing to the common front (the Pacific) shared by these powers, he remarked (July 6, 1853): "On this great field of action we are promised, amongst the gladiatorial operations, one show which no power short of inspiration could have foreseen." He based this prognostication once more on a demographic fact (August 25, 1853): "The world ought to know, if it does not, that the natural human increase of Russia only falls behind that of the United States." A century before our modern Cold War, Darby calculated a collision course for the two emerging powers.

The prospect of war commanded the writer's attention in what proved to be his final contributions to the *Intelligencer*. Commenting on an impending invasion of Russia by an alliance of European nations, he observed (April 22, 1854): "Once commence the drama and a general war must ensue. It is doubtful whether the world was ever at any time in more imminent danger of one of those deranging convulsions which it seems no human foresight can prevent or control." A month later (May 30, 1854), he argued in a brief note that Russia, like the American colonies, would rise to the challenge of meeting a numerically superior force because her own soil was threatened. With that unremarkable opinion, Darby's quarter-century relationship with the Washington newspaper came to a close.

Darby's health failed rapidly in the summer of 1854; when the Senate passed a bill for his financial relief on July 7, the report described him as "far advanced in life, and in very infirm health."[46] He was no doubt gratified by the $1,500 allocation of Congress which brightened his seventy-ninth birthday on August 14, but fate had allowed him little opportunity to enjoy the recompense. He lapsed into unconsciousness on the afternoon of October 7 and ended his earthly travels on October 9, 1854, at eleven o'clock in the morning. In a letter to Draper, Mary Tanner Darby reported the death of her husband:

> I suppose by this time you have seen by the Papers, the irreparable loss I have met with since the receipt of your last Letter. Mr. Darby is no more, he calmly went to sleep on Monday the 9th Oct. after an illness of six days. he was insencible from Saturday about two O'Clock at which time he told me he was perfectly happy. you know not what consolation those words have poured into my Heart, as I had had some doubts on the subject but the earnest manner and the perfect tranquility afterwards left no doubt on my mind that Mr. D. has gone to a happy place of rest.[47]

Funeral services for William Darby took place at Mrs. Clare's boardinghouse in Washington on the afternoon of October 10.

Gales and Seaton published a brief but sincere tribute to their old friend (October 10, 1854), whose learned and sometimes pedantic essays had graced the *Intelligencer* for so many years. They traced his development from his frontier childhood:

> Reared in that country, he grew with its growth, and, aided by his love of physical and especially of geographical science, he was better acquainted with the geography and history of the Great West than any man we have known. His knowledge was not, however, confined to his own country, but ranged through all the world and

46. U.S. Cong., Senate, *Reports of Committees*, 33rd Cong., 1st sess. (Washington, 1854), II, Rept. 222.

47. Mary T. Darby to Lyman C. Draper, October 18, 1854, in Draper Manuscript Collection.

through all recorded history. These acquirements, engrafted on a mind of remarkable vigor and power of analysis, rendered him the most accurate historian, geographer, and statistician of whom we have ever had any knowledge. Nor was he less remarkable for the wisdom which he drew from the lessons of history and experience. He was a man of singular sincerity, probity, and benevolence, and was especially deserving of respect and of admiration for the powers of his enlightened understanding.

Allowing for the rhetorical inflation of eulogies, one nevertheless discovers here a fair assessment of Darby's attainments at the time of his death. He was indeed, as Gales and Seaton asserted, a remarkable man.

Yet with the death of those who had known him or his writings, Darby's contributions to American geography, history, and belles-lettres virtually faded from human memory. Only brief sketches in reference volumes have preserved his name from total oblivion, and even these have reflected a fragmentary understanding of his life and works. From one perspective, such neglect seems inevitable; Darby never wrote a truly memorable book, nor did he exert much influence on younger geographers and historians (Draper, for example, seems not to have paid much attention to Darby's exhortations). Superficially regarded, his published works seem the effusions of a harmless drudge given to statistical pedantry. Furthermore, a good deal of his thought and research found expression in ephemeral forms—lectures, newspapers, magazines, and gazetteers—which were soon outdated or forgotten. And thus despite the esteem of Gales and Seaton, the achievements of William Darby slipped into deepening obscurity in the years following his death.

Commenting on the vagaries of historiography, Darby once remarked, "I have seen, heard, and read enough of distortion in regard to what I knew to be real in action, to doubt much of all history. So many examples of the worthless exalted and of the deserving trampled underfoot, that confidence in reality cannot be preserved." [48] As we have observed in his letters to Draper, the geographer believed himself to be one of those deserving of a

48. William Darby to Lyman C. Draper, August 15, 1850, in Draper Manuscript Collection.

place in the history of the West. And so he does, but for reasons quite different from those which Darby might have offered. His claim to our attention does not rest entirely upon his geographical writings (though his study of Louisiana contains important, useful material) nor his travel sketches (despite the readability of his *A Tour from the City of New-York, to Detroit*), nor upon the border tales of Mark Bancroft; his significance derives as much from the example of his life. Darby was in several respects a prototypical figure, a man of intelligence and literary ambition who dreamed one of the compelling dreams of his age: to know the American West and to achieve fullness of experience in that cultivated garden.

For all of his plodding, meticulous compilations, Darby was a man of vision, driven by a passion to make known the history and resources of the West—not for the sake of information alone, but to encourage the great revolution in American life which absorbed his generation. He conceived of westward movement as a necessary, ennobling enterprise, which confirmed the worth of Anglo-Saxon civilization as it transformed the savage wilderness. The alluring myth of the garden of the West determined the course of his early years, first in his family's migration to western Pennsylvania and then in his long journey to the Old Southwest. And despite the bitter failures of the latter adventure, the West continued to dominate his thoughts; during the last half of his life, Darby's periodic excursions from the cities of the East to the pastoral haunts of his youth took on an almost ritualistic aspect.

This intriguing pattern of cyclic return began in the fall of 1815 and continued through his 1852 visit to Columbus, Ohio. His travels were motivated in part by practical concerns: the gathering of geographical data, the pursuit of employment, the need to recuperate from the griefs and illnesses of his last years. Perhaps all of Darby's visits to the West were further prompted by an unconscious desire to go home again to the place of his frontier childhood, to seek out those dwellings to which he returned in dream and memory. But the essential reason for his revisiting the West seems bound up with his compulsive interest in change: he returned to the border country to experience the delight and astonishment which evidence of human progress invariably excited.

By returning periodically, Darby witnessed the transformation of "the Western country" as a series of juxtaposed images, each reflecting a more advanced stage of physical and cultural improvement. Darby perceived this change much as Thomas Cole had portrayed it in his allegorical series "The Course of Empire"; yet unlike the painter, who visualized final decay, the geographer observed only the "rising civilization" of the West. Despite his inveterate habit of regarding history as a lesson in the rise and fall of nations and despite his occasional concern for the depopulation of the East, Darby found the decline of America literally unthinkable. What he saw in the West inspired only boundless optimism, as if the unique conditions which generated the myth of the cultivated garden also insured the nation against the ordinary catastrophes of history.

The meaning of Darby's varied career seems ultimately rooted in this integral relationship between his life and his writings. The diagram of his travels is, in a sense, an objectification of those desires and beliefs which inform his prose; his long journey to the Southwest and his repeated visits to the Pennsylvania-Ohio frontier mirror his intellectual curiosity, his quenchless determination to witness and describe the glorious civilizing of the West. Darby was an authentic pioneer—a settler on the Pennsylvania border, an explorer in Louisiana, a bellwether of American geography, and an early exponent of statistical methods. By dint of personal effort, he transformed himself from an ignorant backwoods youth into a creditable scholar and scientist—a civilized man—thus effecting in his own life the kind of amelioration which he delighted to see in frontier communities. Though his interests were diverse, no subject engaged his mind and heart more completely than "the Western country." He strove in his early authorial efforts to compile accurate information about the West for an often misguided public; he endeavored to clarify and preserve western history in the narratives and sketches of his middle years; and he turned in his later years to speculative essays on population growth and the culmination of westward movement. Though long overlooked by modern historians, Darby's story bears telling, for it enables us to understand in personal terms the wonder and passion which the opening of the West once inspired.

# II. The Magazine Fiction

# Mark Bancroft's Border Narratives

In William Darby's many-faceted career as a man of letters, no other chapter reveals more about his personal concerns or the literary life of that era than his eight-year engagement as a magazine writer. He had been introduced to the hard facts of periodical publishing, of course, in 1824, when his *Monthly Geographical, Historical, and Statistical Repository* failed after the second issue. But the appearance of the *Repository* coincided with a pivotal development in American literature and culture, one destined to lure Darby back to the public prints: that "extraordinary outburst of magazine activity" in the mid-1820s, which produced a sudden demand for literary effusions of all kinds.[1] The magnitude of this phenomenon can be suggested statistically: the number of American periodicals (excluding newspapers) increased sixfold between 1825 and 1850.[2] Scientific, commercial, and religious journals sprang up with amazing rapidity; literary gazettes and belletristic magazines likewise proliferated. The latter fact has a special importance: at a time when the absence of an international copyright law stifled the publication of books by Americans, periodicals offered an essential outlet for native authors and a means of cultivating a truly national literature. Among other results, the magazine movement

1. Frank Luther Mott, *A History of American Magazines* (Cambridge, Mass., 1939), I, 340–41.
2. *Ibid.*, 341–42.

gave rise to the American short story and elicited the first tales of such writers as Poe and Hawthorne.

One of the more consequential journals to appear during the mid-twenties was Samuel C. Atkinson's *Casket*, the forerunner of *Graham's Magazine*. Founded in 1826 as the monthly counterpart to the *Saturday Evening Post* (inaugurated three years earlier), the *Casket* met with commercial success; one contemporary writer called it "the most widely circulated monthly" in the United States.[3] Though its exact circulation is unknown, the accretion of sales agents over a short period gives some index of its popularity: in April, 1831, the magazine listed 176 representatives, but by August, 1833, it boasted over 400 in twenty-seven states and territories. Much of the magazine's appeal lay in its diversity. In addition to fine engravings and tinted fashion plates, tales pirated from British journals, songs (with music), and a potpourri of jokes, riddles, and anecdotes, Atkinson's *Casket* featured scientific articles by C. S. Rafinesque, sentimental tales by Eliza Leslie and Mrs. E. F. Ellet, graveyard poetry by the "Milford Bard" (John Lofland), satirical pieces by Lambert Wilmer, and, beginning in July, 1829, frontier narratives and historical essays by one "Mark Bancroft." During the journal's most prosperous years, in fact, Mark Bancroft (William Darby) was one of its literary mainstays.

Oddly enough, Darby's acquaintance with Samuel C. Atkinson probably began with an erroneous notice of Darby's death in the *Post* (August 18, 1827). However disconcerting, the obituary lavished praise: "Mr. Darby as a topographical engineer, had few if any equals, and as a geographer, he was never excelled in modern days. His decease is a severe loss to science in general." Darby's wry response—which calls to mind Mark Twain's protest that reports of his death had been "greatly exaggerated"—was published in the *Post* (September 4, 1827) and deserves reprinting here:

> Sandy Springs
> Montgomery County, Md.
> August 24, 1827

Messrs. Atkinson & Alexander,
    I this day for the first time, saw in your paper of the 18th instant,

---

3. Lambert A. Wilmer, *Our Press Gang* (Philadelphia, 1859), 19.

a notice of the death of William Darby, on the 30th ult. in Frederick county. I had been previously informed by letter from Henry S. Tanner, that such a report was in circulation. Before I actually read the account, I had thought that some unworthy design had prompted so very unfounded a report; but the friendly tenor of the notice has convinced me that the whole affair has originated from some unexplained mistake. I have consulted some of my neighbors, but none has attempted to unravel the mystery. There are in this county several families of the name of Darby, but no report of a recent death amongst them has been published to my knowledge.

As to myself, my life has not flowed in so unruffled a current as to render its continuance a subject of much anxiety, but as I have a family, many sins to repent of, and some infirmities to amend, as well as much projected duty to perform, I would, if I had my own choice, prefer living a little longer. I am now engaged in a "Philosophical View of the United States," and have in some preparation material for a future edition of my Geographical Dictionary. In brief, I have sufficient work cut out for ten years assiduous labor. Such a term will bring the eve of the grand climacteric, when I hope to be more resigned, and better prepared than I am at present, to *leave the warm precincts of the cheerful day*.

William Darby

Less than two years later, Darby had become a regular contributor to Atkinson's periodicals. His earliest tale, "The Vendue," appeared in the weekly in four installments during June and July, 1829, and the *Post* for October 10 featured "The Sioux Chief," a narrative of border warfare drawn from Darby's personal acquaintance with hunter-warriors of that epoch. In an accompanying editorial note Atkinson puffed the story and hinted at subsequent works in the same vein: "This tale . . . is from the pen of a gentleman who has resided for many years in the immediate neighborhood of the Indian tribes, whose habits of observation peculiarly qualify him for the task of description. It will most probably be the commencement of a regular series of narratives on this subject, the leading object of which will be to render the manners of the Indians more familiar than they have hitherto been, conveying this information through the medium of pleasant fictions, which will thus combine interest with amusement." Clearly, the Philadelphia publisher considered Darby an authentic frontiersman, versed in the ways of Indian life; Atkinson doubtless hoped to tap the kind of popular interest that made Schoolcraft's narratives so widely reprinted in the magazines of the period. Yet after the publication of a sequel to "The Sioux Chief" titled "The Indian Trader; or, James Bolton, of Orange" (*Casket*, December 1829), Darby drifted toward oriental tales and conventional romances—works having little to do with frontier life or Indian ways. From 1830 until mid-1834 he contributed more than a dozen narratives to Atkinson's journals, but only two dealt concertedly with the history of border warfare: "The Hunter's Tale; or, Conrad Mayer and Susan Gray" and "The Moravian Indians." To these we might add an unsigned tale, quite possibly by Darby, titled "Washaloo, the Indian Sachem; or Faith Unbroken" (*Casket*, September 1833). On the evidence of his published tales, Darby the fictionist seems to have been pulled in two directions: his pedagogical instincts prompted him to conceive of the narrative as an aid to historical understanding, while his literary aspirations inclined him toward all of the sentimental conventions which then governed periodical writing.

During the eight years in which Darby supplied material to Atkinson, the nature of his contributions varied widely. In addition to nearly thirty full-length narratives, he published numerous essays on European and American history and several travel sketches. But among his writings, the Philadelphia publisher especially prized Darby's frontier tales, and in March, 1834, he offered a "regular engagement" to supply narratives and essays on western life, at the rate of $1.50 per magazine page.[4] The first of these, "Mark Lee's Narrative," reached Atkinson in mid-April and inspired more editorial fanfare in the *Post*: "We thank our friend Bancroft for his excellent tale, 'Mark Lee,' being the first in a series of Border dramas, touching on prominent incidents connected with American history. We have no doubt the series will be properly appreciated by all who have read the 'Vendue,' and other tales from the untiring pen of our correspondent Mark, and to those who have not experienced that gratification, we can safely promise

---

4. *Notes and Queries* letter, 5. Darby's rate of pay is mentioned in a letter to D. Christy, February 28, 1839, Manuscript Division, New-York Historical Society, New York, N.Y.

a rich intellectual treat on the proposed articles."[5] The subsequent "Border dramas" included some of Darby's better historical studies—"Cyrus Lindslay and Ella Moore" (*Saturday Evening Post*, June 20, 1834), "Gilbert and His Family," and "The Wedding." Through Atkinson's encouragement, the geographer published fourteen tales in the space of two years, nearly all concerned with the vicissitudes of frontier experience or the facts of western history.

Darby began to write magazine tales at the height of his productivity as a man of letters; he was then in his fifty-fourth year. Having published three recent books on geography or history, he perhaps turned to prose narratives in the spring of 1829 as a diversion from his other work.[6] But he may also have wished to satisfy certain literary ambitions; his tales seem partially inspired by the success of James Fenimore Cooper, whose frontier hero, Leatherstocking, Darby once described as "one of the most perfectly natural characters ever sketched."[7] The narratives may have been prompted in part by a historical motive, for in many of them he strove to depict those personages and events reflective of pioneer times and to re-create the ambience of that period. His didactic tendencies compelled him to insert factual information about the border wars and Indian life, while nostalgia induced him to dramatize scenes from his own boyhood. In several ways the tales enabled him to examine and clarify the meaning of the westward movement in his own life and in the life of the nation.

If they expressed certain personal concerns, Darby's narratives also embodied many of the conventions of popular magazine fiction. We shall pause here only to note some conventional aspects of style, one of which was his predilection for effusive language. Here and there he rivaled Cooper in descriptive excess; the opening lines of "Lydia Ashbaugh, the Witch" (*Saturday Evening Post*, January 16, 1836) exemplify the use of inflated rhetoric: "Though the Appalachian steeps do not rise to Alpine heights, nor do they

5. *Saturday Evening Post*, April 19, 1834.
6. In 1828/29 Darby published *A View of the United States, Lectures on the Discovery of America*, and *Mnemonika; or, The Tablet of Memory*. He was also preparing an edition of the *United States Reader*.
7. "Reminiscences of the West," *Casket* (December, 1834), 542*n*.

aspire to vie with the towering Cordelleras, still they rise rock upon rock, wood crowned to awaken feelings of admiration and grandeur in the bosom which swells upon their rocky sides or frowning brows. In infant years I gazed upon these fringed dells and beetling cliffs. . . ." Like many of his contemporaries, he injected literary allusions and quotations into his tales, and he regularly employed poetic headnotes, culled from the works of Shakespeare, Milton, Wordsworth, Coleridge, Byron, Shelley, Campbell, and Felicia Hemans. He apparently had a special regard for Waller's "Go, Lovely Rose," for he frequently described the heroine of his tales as "a flower in our wilderness" or "a rose on the desert, born to blush unseen." These stylistic traits seem of secondary importance, though; other aspects of Darby's tales—his narrative persona, his adaptation of historical material, and his thematic derivations from the adventure tale and the tale of sentiment—deserve more careful examination.

With his first published narrative, "The Vendue," Darby established the fictional guise of Mark Bancroft, a mask he retained throughout his career as a magazinist. Though generally recognized, after 1843, as the author of the Tacitus essays in the *National Intelligencer*, Darby concealed his identity as the author of the Mark Bancroft tales so successfully that when Lyman C. Draper first corresponded with him in the summer of 1845, the historian had asked naïvely, "Some years since, there was a writer who wrote over the signature of Mark Bancroft—who was he? and is he yet living?" The question "excited a smile" from Darby, who replied: "This query I can answer fully, Mark Bancroft is, I assure you living and well, though only 13 days short of 70 years. He was born August 14th, 1775, and now has the pleasure and honor to introduce himself in his proper name of William Darby."[8] The exact motives for the pseudonym remain a matter of conjecture, though the practice was then common enough in periodicals. Perhaps Darby's innate diffidence prompted the mask, or perhaps he felt the persona granted special privileges in the use of biographical or autobiographical material. Whatever his reasons, the genial storyteller Mark Bancroft gradually acquired recognizability, be-

8. William Darby to Lyman C. Draper, August 1, 1845, in Draper Manuscript Collection.

coming a discrete personality similar but not identical to Darby himself.

To be sure, in some of the border tales the narrator is either effaced (undramatized) or virtually indistinguishable from the author. But in other stories, Bancroft enters into the action or speaks of himself, permitting us to glimpse intriguing differences between himself and his creator. When we first meet him in "The Vendue," he is portrayed as an elderly bachelor, "unconnected with any human being in America, by either blood or marriage." In that tale we see him purchasing land in 1804 on the Muskingum River in Ohio and residing tranquilly for over twenty-five years amid the Swansey and Overton families (whom we meet in "The Vendue"). That is, Mark Bancroft lives the settled life of a bachelor which Darby himself never experienced. Only one tale—"Ann Eliza Glenn, a Tale of Wyoming" (*Saturday Evening Post*, September 12, 1835)—represents him as a married man, traveling in the Wyoming Valley of Pennsylvania with "the partner of all my wanderings." Characteristically, though, the narrator appears as a solitary wanderer, making his way toward some obscure destination. For example, the opening paragraph of "Clement Meyerfield and Clara Ismeana" (*Casket*, May, June, 1830) finds the narrator in Louisiana, on "one of those extended pedestrian journies" which he has made "in the wide spreading west." Similarly, in "Lydia Ashbaugh, the Witch" the narrator sets the scene of his tale by remarking, "In one of my rambling excursions I rose a mountain path but little frequented in the northern part of Franklin county, Pennsylvania." Though Elizabeth Darby did in fact accompany her husband on many of his "rambling excursions," the author chose to portray Mark Bancroft as a solitary figure, probably for romantic effect and technical convenience. What has been suggested about Cooper's Natty Bumppo may also be true for Darby's narrator: the foot-loose frontiersman may be a projection of the author's submerged desire to escape from the constraints of domesticity.[9]

Those tales that employ a dramatized narrator typically begin

9. Leslie Fiedler sees Natty Bumppo as "a dangerous symbol of Cooper's secret protest against the gentle tyranny of home and woman." See *Love and Death in the American Novel* (New York, 1966), 194.

with a chance encounter: Mark Bancroft meets another individual in a lonely or remote place or he comes upon a house or tavern and there makes an acquaintance. In either case, he becomes a sympathetic listener who records the story told to him by some denizen of the backwoods. Darby uses this favorite framing device in "Ellery Truman and Emily Raymond, or The Soldier's Tale," when the narrator, pausing to admire the landscape, suddenly confronts a stranger: "A slight rustle on my right hand recalled me from my reverie, and turning quickly toward the sound, met the eye of an aged man, who had been very attentively watching my abstraction." The old man, a veteran of the Revolution, accompanies Bancroft to Derry, and there in the churchyard, over the grave of the Reverend John Roan, tells his poignant story of romance and battle. A variation of this device occurs in "Mark Lee's Narrative": having stopped at an inn on their westward journey, the narrator's family meet a frontier scout who apprises them of the real danger of western life—the white man's foolishness—by relating several anecdotes. In "The Hunter's Tale; or, Conrad Mayer and Susan Gray," Bancroft chances to discover an old acquaintance, Kingsly Hale, and listens as Hale spins a story about the fabled Lewis Wetzel to a group of young people gathered on the banks of the Ohio. Through such encounters, Darby's narrator thus comes to be seen as a peripatetic collector of tales, a backwoods historian, whose amiable, unassuming manner wins the confidence of everyone he meets. Only once, in "The Unknown" (*National Atlas*, July 31, 1836), does he comment explicitly upon his literary aims: "My principal object in advancing towards seventy, was to record the fates and fortunes of obscure persons like myself, and thus aid in doing the small work which was neglected by the proud genius of history." Like the speaker in Gray's "Elegy Written in a Country Church-yard" (a poem Darby clearly admired), Mark Bancroft regards himself as the memorialist of "obscure persons" whose "fates and fortunes" possess human interest and comprise the bedrock of common experience which history typically ignores.

As a wanderer guided only by this broad, historical objective, Mark Bancroft may be readily distinguished from the laborious statistician and geographer who created him. Yet it is clear too that the narratives contain a good deal of autobiography and collectively mirror the course of Darby's travels over a half-century. If Mark Bancroft is a distinctly fictional persona, he has much in common with the author: "Mark Lee's Narrative" and "Reminiscences of the West" both recount an arduous journey to "the Western country" in 1781; "The Wedding" presents the narrator as a young man in Washington, Pennsylvania, in 1791; "Gilbert and His Family" depicts both a visit to the Lehigh River in 1821 and an earlier journey along the Youghiogheny River in 1796. To these unmistakably autobiographical episodes can be added other details and scenes presumably drawn from Darby's travels. His unpleasant river voyage from Pittsburgh to Natchez in 1799 surely contributed to "Ann Dillon" (*Saturday Evening Post*, January 9, 1830), in which the heroine tells of the oppressive heat on the Mississippi and the myriads of stinging insects swarming about the riverboat in early August. "Caroline Marlow" includes vignettes from Darby's residence in Natchez, while his sojourn in Louisiana enabled him to add realistic touches to "Henry and William Nelson" (*Saturday Evening Post*, November 13, 1830), such as this scene along Bayou Teche: "It was on one of these excursions, on a sweet autumn evening, that Henry was slowly pacing the road, admiring the sweep of prospect towards the setting sun, and contrasting the fading leaves of the other forest trees along the Teche, with the deep green of the live oaks and laurel magnolias, when the carriage of James Sevier whirled past." In the same manner, Darby's 1818 journey along the St. Lawrence River inspired the picturesque setting of "The Shipwreck" (*Casket*, April 1831): "The shores are walls of rock, rising abruptly from the almost pellucid water into precipices crowned with wood or farms. The whole region, in a summer's sun, is worth a voyage." Tales such as "Ann Eliza Glenn" or "Lydia Ashbaugh, the Witch" show Darby's familiarity with the mountainous interior of Pennsylvania, while "The Spirit of the Potomac" (*Saturday Evening Post*, December 18, 25, 1830) sketches the Maryland countryside he knew from his rustication in Montgomery County.

But while the narratives display the extent of Darby's travels, his interest as a writer invariably lay with characters and their personal circumstances rather than physical setting. Bancroft sometimes notes differences of speech or custom, but "local color" matters less to him than the patterns of human action found everywhere. This generalizing habit is understandable when we remember that the local color movement of the later nineteenth century sprang from the provincialism—that is, the vivid but narrow experience—of its participants. Darby's travels acquainted him with the conditions of pioneer life from Pennsylvania to Kentucky to Louisiana; though conscious of cultural or regional characteristics, he was always more fascinated by the broader distinctions of wilderness, frontier, and civilization. Thus in Mark Bancroft's tales the movement from the city to the backwoods is of far greater thematic importance than the movement from one western state to another. Though the author would perhaps have resisted such a dichotomy, we might say that the historian prevails over the geographer in Darby's border tales—that the human history of frontier life takes precedence over the specific features of climate, terrain, and land use which conditioned that experience.

Mark Bancroft explains in "The Unknown" that he aimed to record "the fates and fortunes of obscure persons" neglected by history, but that remark accounts only partially for the substance of Darby's narratives. In fact, Darby's desire to portray typical or representative incidents ("the fates and fortunes") of pioneer life was balanced by an intense interest in the real events and personages of border history. Thus we find in many of the tales a sustained interplay of fiction and fact, a merging of the imaginative and the mimetic. Whatever Darby's limitations as a stylist, we find in his narratives surprisingly diverse uses of historical materials and complex methods of uniting the actual and the imaginary. It is in this regard, perhaps, that Darby's work most fully exhibits the romantic-historical sensibility of his time; the more accomplished art of Scott, Cooper, Simms, and Hawthorne grew from the same combinative impulse. In Darby's treatment of the past—particularly the border wars—we see him adapting a popular literary mode to suit his needs as a historian of "the Heroic Age."

Darby expressed the truth when he wrote to Draper in 1845: "I take a deep interest in whatever relates to the Great West. . . . The West is the home of my imagination and warmest recollections and aspirations."[10] In addition to heartfelt memories of his frontier youth, though, he retained impressions of fearful events: the flight to Jacob Wolf's fort after the massacres of early 1782; the Washington County militia departing on bloody expeditions under Williamson and Crawford in the same year; the murder of the Crow girls and the Becham boys in 1787; the slaughter of the Tush family in 1794; and the killing of two Indians near Wheeling in 1795. As a young man, Darby's curiosity prompted him to seek fuller information about these and other episodes from pioneers directly involved in Indian fighting. Thus during his stay in Wheeling, he gained inside knowledge about the siege of 1777 from such veterans as Henry Jolly, Lewis Bonnett, Sr., and Jonathan Zane. Explaining the vividness of his frontier recollections to Draper, Darby admitted: "As to myself, what I have stated of my age when brought on the Great Western Scene, you will of course see I could be only a spectator, and a very young one also, of the most trying scenes. . . . But on the other hand, mixing with the actors personally, whilst the events were recent, and memory active, I seem now to see their faces and hear their voices."[11] If the accounts of these "actors" shaped Darby's early conceptions of frontier history, a major influence in later life was apparently Alexander S. Withers' *Chronicles of Border Warfare*; after its publication in 1831, the work became a veritable touchstone, which he cited copiously in such pieces as "Mark Lee's Narrative" and "Reminiscences of the West." Given Darby's voracity as a reader, we must assume that he was also well acquainted with such contemporary historians as John Heckewelder (whom he cites in "The Moravian Indians"), Joseph Doddridge, and Timothy Flint. These authorities provided a broader view of the period as a whole, placing in perspective those events that kindled his imagination.

Drawn from personal experience, oral testimony, and published scholarship, Darby's knowledge of western history ripened into a literary passion in 1829. One precipitating factor may have been Andrew Jackson's victory in 1828. Darby's political differences with Jackson aside, the elevation of a backwoodsman to the presidency conferred a special distinction on frontier life and height-

10. Darby to Draper, August 1, 1845, in Draper Manuscript Collection.
11. *Ibid.*

ened popular interest in the history and traditions of the West. Perhaps more conscious of his own frontier origins as a result, Darby embarked on a series of border tales designed to celebrate western life and to re-create "that age of poverty, heroism, and inflexible purpose."[12] He also intended, from the beginning, to illuminate a vexed and important subject: the turbulent relationship between the white man and the Indian. The *Saturday Evening Post* for October 10, 1829, described "The Sioux Chief" (his earliest tale to use historical material) as the first of a series depicting "the manners of the Indian." But as that tale makes clear, Darby also planned to treat the "manners" of the frontiersman and to depict the extremes of heroism and villainy exhibited by both races. Though convinced that a "moral force" (civilization) impelled the white man's invasion and conquest of Indian lands, Darby nonetheless respected the native American and attempted a just representation of his social and intellectual nature. Rather than portray a simplistic and historically inaccurate version of the border wars—the hunter-warrior prevailing over the cunning savage—Darby found himself obligated to a more truthful reconstruction of the past. By examining a handful of tales, we can perceive the extent of his commitment to historicity as well as his inventiveness in adapting historical data to the demands of narrative.[13]

The most thoroughly historical of Darby's tales is "Gilbert and His Family," a documentary account of a Quaker family's captivity by the Indians in 1780. In a footnote Mark Bancroft comments on its nature: "The readers of the Casket and Evening Post, in the habit of seeing my signature to Tales, which, though generally founded on fact, contained much of pure fiction, may very naturally regard this in the same light; but such, however, it is not. I was intimate with a part of the Gilbert family, and the story is

12. *National Intelligencer*, May 30, 1835.

13. Darby's own statements on the historical reliability of his narratives are somewhat contradictory. He remarked in the *National Intelligencer*, May 30, 1835: "Some of the readers of the *National Intelligencer* will no doubt recollect to have seen in the Philadelphia *Saturday Evening Post* some tales over the signature of Mark Bancroft. . . . Those tales were written by me, and are generally founded on historical facts." Yet eleven years later, on July 24, 1846, he wrote to Lyman C. Draper: "When I wrote the Tales for the *Casket* under the signature of Mark Bancroft, they were written as Tales, and not as History, except 'The Moravian Indians'" (Draper Manuscript Collection).

from their own document, and I very confidently send it forth as truth." Darby had in fact excerpted most of the tale from the pamphlet written by his friend and mentor, Benjamin Gilbert, Jr. With evident respect for the factual integrity of Gilbert's document, the geographer declined to embellish the narrative with a fictional subplot or to shape it for dramatic effect. Yet the piece possesses intrinsic interest, particularly insofar as it portrays the endurance of the white prisoners and the customs of their Indian abductors. Like most captivity narratives, "Gilbert and His Family" emphasizes the Indian's ability to "quench any latent spark of humanity"; yet for all the hardships imposed on the Quaker family, only one of the fifteen captives died during the two-year ordeal. With some justification, Darby concluded: "The final safe return of the whole of Gilbert's large family to the place of their nativity, stands certainly unparalleled in the long history of Indian warfare."

In contrast to the scrupulous historicity of "Gilbert and His Family," two other tales—"The Sioux Chief" and "The Moravian Indians"—illustrate Darby's effort to develop a fictional plot in conjunction with a historical account. In the prefaces of both works, the author noted their factual basis: "The Sioux Chief" depicted the 1780 siege at Fort St. Louis, the killing of an Indian leader, and the intervention of General George Rogers Clark;[14] "The Moravian Indians" re-created the slaughter of nearly one hundred peace-loving Indians by Colonel David Williamson's militia regiment in 1782. Darby had, moreover, heard eyewitness accounts of both episodes. Simon Burney and Thomas McKim, frontiersmen who figure importantly in "The Sioux Chief," actually participated in the siege and related their adventures to Darby in Louisiana many years later. McKim claimed, in fact, to have shot the Sioux leader himself, and he showed the geographer a "long smooth bore gun" with which he had supposedly per-

---

14. Darby's account raises two historical questions: one is his portrayal of Clark's role and the other is the identity of the slain Indian leader. Clark probably took no active part in the siege (contrary to Darby's view), though his presence in the area apparently discouraged the Indian attackers. Darby depicts the killing of the Sioux chief, Pied de Renard; the real Sioux chief at St. Louis, Wabasha (or Wapasha), survived the conflict and died of old age. See John R. Spears, *A History of the Mississippi Valley* (New York, 1903), 318, and Doane Robinson, *A History of the Dakota or Sioux Indians* (1904; rpt. Minneapolis, 1956), 120.

formed the deed.[15] Among other sources, Darby gleaned details of the Moravian massacre directly from Colonel Williamson, a family friend whose reputation the author hoped to vindicate. Yet upon both this tale and "The Sioux Chief," Darby saw fit to superimpose wholly fictional plots, calculated to excite sentimental interest. Thus in "The Moravian Indians" we find the romantic story of Saul Garvin, a lovelorn white man who joins the Moravian tribe, marries an Indian woman, and then, after sending his wife to safety, dies in the massacre with his adopted Red brothers, refusing to acknowledge his white skin even to save his life. To "The Sioux Chief" Darby attached the ironic plot involving Thomas McKim, his brother Leonard, and the Indian chief, Pied de Renard: Thomas mortally wounds de Renard, believing the latter to be his brother's murderer when in fact the Indian had saved Leonard's life. To heighten the poignancy of de Renard's death, Darby has Leonard expatiate on the senseless killing of his friend and recount the equally pathetic tale of de Renard's struggle to forgive the racial bigotry of the white man. Thus in both narratives, the author reconstructs an actual incident from border history that serves not as the mere background but as the determining context of fictional action.

A somewhat different mix of fact and fiction occurs in "The Hunter's Tale; or, Conrad Mayer and Susan Gray" and "Cyrus Lindslay and Ella Moore," both of which portray an authentic historical personage in an essentially fictitious adventure. Each depicts the prowess of a real hunter-warrior, whose fabled skills enabled a younger companion to rescue his sweetheart from Indian captivity. Clearly, the author contrived "The Hunter's Tale" chiefly to display the woodcraft of Lewis Wetzel, the legendary Wheeling scout who possessed the singular gift of being able to reload a flintlock while "running through the forest, pursued by an Indian enemy."[16] The specific events of the tale—Wetzel's encounter with Conrad Mayer, their dispute over a dead deer, the massacre of the Mayer family in Conrad's absence, and the ab-

15. William Darby to Lyman C. Draper, July 22, 1846, in Draper Manuscript Collection. McKim's story, accepted at face value by Darby, does not score with Robinson's information (note 14). McKim perhaps mistook the rank of his victim.

16. A biographical sketch of Lewis Wetzel appears in De Hass, *History of the Early Settlement*, 344–65.

duction of Susan Gray—sprang from the writer's imagination, though he allowed that similar incidents "in some case or other really happened within forty miles of Wheeling between 1770 and 1795." Two key scenes dramatize Wetzel's talent: he resolves the hunter's quarrel by exhibiting his ability to shoot and reload on the run, and he performs the feat later in the rescue sequence to dispatch an Indian assailant. His gallantry produces significant results, bringing together the separated lovers, avenging the slaughter of the Mayer family, and confirming the crucial role of the hunter-warrior in the defense of western settlements. The use of a historical figure to validate a fictional plot proved so convenient that Darby followed the scheme later in "Cyrus Lindslay and Ella Moore," which describes Lindslay's deliverance of the captive Ella with the assistance of Daniel Boone. The author probably derived certain details from Boone's celebrated 1776 rescue of his daughter and the two Calloway girls,[17] but the sentimentalized climax—in which the wounded Lindslay loses an arm but gains a wife—completes a purely fictional love story. Though Boone dominates this story less completely than does Wetzel in "The Hunter's Tale," his extraordinary ability to follow a trail saves Ella from the proverbial fate-worse-than-death, and like Wetzel, he appears slightly larger than life in Darby's historically grounded portrait.

A fourth means of uniting history and fiction manifests itself in the author's 1836 tale "The Wedding." Here, as in other works, Darby locates his narrative in a specific time and place but refers to contemporary events only to sketch a distant backdrop for the central action. To be sure, a historical personage—the Reverend John McMillan—appears in the tale, and the main complication of the plot stems from the disappearance of Powers Osborne in Crawford's fatal militia campaign of 1782. But McMillan simply fulfills a technical requirement (a wedding must have a minister), and the history of Crawford's defeat matters only insofar as it accounts for the absence of Anna Osborne's husband. The story does, however, abound in historical allusions and generalizations about the period. Recalling his boyhood, Mark Bancroft describes the year 1781 as "a time of painful but heroic suffering" in western

17. See the *Notes and Queries* letter, 39, and also M. L. Dixon's account of that episode in the *National Intelligencer*, May 30, 1835.

Pennsylvania, and he contemplates the "fearful and dark tragedy" of the border wars, tracing the events of 1780–1782 that culminated in the rout which supposedly cost Powers Osborne his life. He indirectly contrasts that fateful period with the conditions in Washington County at the time of the narrative (1791): "In the face of every danger and every privation, farms had spread, villages began to rise, schools had been formed, and places of worship had been erected." In short, the area exhibited that happy transformation from savage frontier to cultivated garden which so delighted the geographer. Yet after the fictional denouement—the dramatic reappearance of Powers Osborne, who stops the marriage of his "widow" to his old companion, Matthew Johnson—Mark Bancroft imposes another historical perspective by reminding us that the tender reunion at the close of "The Wedding" took place many years ago, since which time the border country has undergone yet another change through depopulation. The decaying Osborne cottage symbolizes the mutability of the idyllic cultivated garden:

> Drawn away by that infatuation which places paradise on the outer verge of civilization, the father, son, and uncle, sold their sweet home, and plunged into the deep west, and became utterly lost, long years lost to all my inquiries, and the last time I passed the Osborne cottage, I found the house disfigured by waste and neglect; the orchard trees were decayed or dead, and the stumps of the fine sugar maple shade trees only stood monuments of barbarism—weeds and briars covered the garden.

The affecting imagery of this scene indicates the author's awareness of a fundamental irony in the frontier experience: westward movement, the advancing wave of civilization itself, also implied an emptying-out of previously settled areas, leaving behind—at least potentially—a wake of desolation.

Using the diverse methods summarized here, Darby attempted to incorporate historical material into his fiction; roughly one-half of his thirty-two tales and sketches contain similar documentary content. In his remaining works, set typically in a pastoral area of the West or South during the period 1775–1825, the author made no substantial use of border history. Yet the latter narratives utilize many of the themes and plot devices found in the former group.

Consideration of these ingredients seems essential, both to provide a broader view of Darby's interests as a writer and to isolate and identify the basic patterns of signification in his work. As we shall see, these inherent designs reveal a major influence on Darby's narratives—literary convention.

One of the crucial distinctions between literary art and popular literature (that is, classic versus ephemeral writing) is the tendency of the latter to proliferate according to formulas which make possible the immediate replication of a specific type. In prose fiction, the Gothic novel probably represents the earliest instance of a popular, formulistic genre; in the space of three decades, hundreds of works, conceived according to the *Castle of Otranto* formula, sprang into being. The magazine fiction of the early nineteenth century, produced rapidly for a mass audience, also tended to satisfy reader expectations within a limited range of formularized possibilities, and this conventionality manifests itself clearly in Darby's narratives. A reading of the entire canon discloses the recurrence of a few recognizable character types, functioning in predictable ways to create those situations required by the author's two principal generic models—the sentimental tale and the adventure tale.

Obviously, the scope of this essay does not allow complete definition of these dominant genres, nor can we develop rigid distinctions between the two, since they could easily intersect in a given work. But we can determine, for purposes of discussion, those formal characteristics that lie behind our intuitive recognition of the types. The sentimental tale, which in a sense produced *Godey's Lady's Book* in 1830, typically hinges on the creation and destruction of human relationships and derives its emotive effect from the interplay of love and death, beauty and disease, or separation and reunion. As its name suggests, this type attempts to generate intense sentiment (usually joy or grief) through scenes transparently contrived for that purpose. The adventure tale, on the other hand, portrays physical movement and conflict, usually a journey that involves struggle against natural forces or human adversaries. Set mainly on the frontier or at sea (during the period in question), this type depicts the male-dominated, outdoor world, in contrast to the female-oriented, domestic world of the sentimental tale. As this dichotomy suggests, editors and authors of the day operated under the assumption that men and women had utterly different literary tastes—a fact borne out by the names of such journals as the *Album and Ladies' Literary Gazette* or *Burton's Gentleman's Magazine*. Female readers, it was popularly believed, preferred the chaste melancholy of the sentimental tale, while male readers yearned for the rough heroics of adventure narratives. Whether these distinctions have any factual basis is beside the point; clearly, they influenced the production of literature in the early nineteenth century.[18]

These differences have relevance to the present study, inasmuch as the *Casket* (like other successful magazines) purposely courted readers of both sexes—and Darby fashioned his narratives accordingly. Adventure themes thus inform about twenty of his tales and sketches; like Cooper (who also combined action and romance), he relished battle scenes, particularly those with some historical authenticity. We have already noted his considerable interest in the border wars, variously dramatized as military conflicts or isolated acts of terrorism. The adventure of battle also figures in his tale of the Revolution ("Ellery Truman and Emily Raymond"), and he elsewhere portrayed the Colombian campaign of Simón Bolívar (in "The Spirit of the Potomac") and Kosciusko's struggle against the Russian army (in "Clement Meyerfield and Clara Ismeana"). Nautical disaster provided another adventure theme, which we find in "The Vendue," in "Ann Dillon," and (most vividly) in "The Shipwreck, or The Hermit of the Rock":

> The stranded vessel and her crew were exposed to all the ravings of the fierce northwest, about sixty yards from an almost perpendicular wall of rock forming the main shore, and at least twice that distance from an island, whose granitic fringe frowned destruction. Between those stony jaws, one wave was vomited rapidly after another, while death was breathed in deepening blasts, rendered peculiarly terrible when commingled with the cries from the vessel and shore.

18. A corroborating discussion of these two narrative modes appears in William C. Spengemann, *The Adventurous Muse: The Poetics of American Fiction, 1789–1900* (New Haven, 1977). Spengemann's opening chapters (pp. 6–118) define "The Poetics of Adventure" and "The Poetics of Domesticity."

Though use of the passive voice undercuts the desired effect, we see Darby here striving to create the sort of violent tensions required by the adventure tale.

Since the *Casket*'s title page promised "Gems of Literature, Wit, and Sentiment," the author also labored to inject sentimental interest into many narratives. Well over half of his tales contain a love story, invariably complicated by parental objections, debt, disease, or war. For example, the father of Maria Edgefield refuses to let the girl marry Pied de Renard, the Sioux chief; Ellery Truman marches off to war and nearly dies of battle wounds before he is reunited with his beloved Emily Raymond; under instructions from her greedy uncle, Anna Elsworth (of "The Unknown") rejects the proposal of impoverished Henry Cordwell. Love affairs also go awry through the chicanery of rivals: in "The Indian Trader; or, James Bolton, of Orange" (*Casket*, December 1829), Bolton publishes a false report of Leonard McKim's death to pressure Alice Letchgrove into an unhappy marriage. Few of Darby's love stories, however, match the mournful quality of "Letburn Parkman, or The Maniac" (*Saturday Evening Post*, January 10, 1835), in which Parkman and his young bride, Lucy Ryland, both enlist for a militia campaign against the Indians in 1781. Disguised as a boy, Lucy is mortally wounded when she tries to prevent her erstwhile suitor, Eli Bringham, from tomahawking an Indian woman; the incident unhinges Letburn Parkman, whose vitality ebbs with his sanity. The two lovers die within a few hours of each other, conveniently making possible a joint funeral.

Nearly as numerous as love stories, though, are those narratives that depict the disappearance and subsequent return of a character—the latter event producing, of course, an emotional and rhetorical outpouring. This pattern occurs in "The Vendue" when William Swansey, unrecognized after an eighteen-year absence, returns to his family and saves them from financial ruin. In "Julia Gray, or The Orphan" (*Saturday Evening Post*, August 2, 1834), the young heroine escapes from a cruel stepmother, assumes a new identity in another state, and then returns to her native village on her eighteenth birthday to claim her dead father's property. As we have seen, this design also appears in "The Wedding," when

Powers Osborne seemingly returns from the dead to prevent his wife's remarriage to another man.

As this overview suggests, Darby's fiction relied heavily upon conventional (and predictable) narrative patterns and reflected little ingenuity in plot construction. Thus reduced to simplest terms, his work betrays its thoroughly formularized nature; throughout the entire range of his tales, in fact, virtually every one may be said to embody some variation of the disaster or separation motif. Disaster typically occurs as overt, physical destruction, though it is frequently accompanied by mental anguish. It comprehends a variety of experiences, ranging from shipwreck and massacre at one violent extreme to financial ruin or debauchery at the other. Separation occurs either as involuntary absence or deliberate abandonment; characters may become separated through adversity, death, or emotional alienation. As Darby uses these motifs, each experience has two potential resolutions: characters may recover or fail to recover from disaster; they may be reunited after separation or fail to achieve reunion. These patterns may be developed separately or combined, usually to connect recovery from disaster with reunion. Simple as this formula may be, it generates nearly all of Darby's narratives and accounts for their outcomes. Leaving aside combinative possibilities, we can readily observe the four basic patterns.

The first, disaster without recovery, determines the action of "Ann Eliza Glenn," a tale depicting the brutal massacre of a Wyoming Valley family. Through sheer chance, the title character and a son survive the atrocity, but Mark Bancroft tells us in the final paragraph: "From the day of her bereavement, Ann Eliza Glenn lived but for her child. The duties of life she performed, but except those of a mother, her feelings seemed dead to human sympathy. . . . From the fatal morning THE WIDOW AND MOTHER WAS NEVER SEEN TO SMILE." But in "The Hunter's Tale" a similar incident culminates in a marriage between Conrad Mayer and Susan Gray, accompanied by the narrator's assurance that "time softened the memory of the past." The disaster-recovery pattern is here complemented by separation-reunion in the abduction and rescue of the heroine. The ubiquitous separation-reunion motif also determines the structure of "Lydia Ashbaugh, the

Witch," in which an old woman sheds a disguise, reveals her identity to an illegitimate daughter, and then saves the daughter and her family from poverty by producing vital legal papers. Once again, reunion coincides with recovery from a desperate situation. But in a tale like "The Moravian Indians" no reunion follows separation: Saul Garvin leaves Ellen Aylworth (his first love) and later sends away his Indian wife in anticipation of the massacre. That disaster precludes both recovery and reunion for Garvin and his wife.

For the sake of brevity, we shall forgo a demonstration of the pervasiveness of these patterns; full examination reveals, though, the consistency with which Darby utilized these four paradigms. It seems unlikely that he applied the formulas in a fully conscious way; rather, he appears to have followed them instinctively and mechanically, disclosing through them, perhaps, the private complications of his own "chequered existence."[19] However, a more verifiable explanation for these insistent patterns emerges when we consider the pressure of literary convention. Clearly, Darby's generic models—the adventure tale and the sentimental tale—required the kind of incidents effectuated by the disaster and separation paradigms, and the compatibility of these formulas allowed him to conjoin adventure and sentiment with relative ease. Disaster provided action; separation supplied poignancy. Out of this coupling came the author's most characteristic tales.

For slightly over eight years, Darby plied his modest skills as a magazine writer. That he lacked grace as a stylist and sometimes stumbled into ungrammatical constructions the modern reader quickly discovers. Conscious of his limitations, he once apologized: "Too much of my life has been spent in actual travelling, to admit the order and polish of a writer who enjoyed more closet leisure; but to relate what I saw, needed not a finished erudition, it demanded a respect for truth and human esteem."[20] Despite their lack of "order and polish," Darby's narratives possess intrinsic in-

19. Darby's biography supplies ample instances of both disaster and separation. His sojourn in the Southwest, for example, involved a cutting of family ties; the ruinous fires of 1804, which ended his career as a cotton planter, produced colossal debts. His surveying excursion in late 1814 took him away from home at the time of his wife's death; that is, he experienced separation without reunion.

20. *A Tour from the City of New-York, to Detroit*, vi.

terest thanks to his "respect for truth and human esteem." As a historian, he endeavored to create truthful accounts of the border wars and frontier life; as a fictionist, he shaped his tales in accordance with prevailing literary modes, as defined in the *Casket*. an original artist, Darby followed a path blazed by such contemporaries as Cooper, Timothy Flint, and James Hall. But he brought to his narratives a rich store of historical knowledge and personal experience, from which he drew to construct his own monument to "the Western country" and its heroes. Through this writing he sought to clarify the meaning of the westward movement as it revealed itself in events, personalities, and attitudes, and he sought further to explore those moments from his own childhood which seemed in retrospect to hold the key to his passion for the West. Ironically, Darby's return to the West in 1836 curtailed his career as a magazinist; though he apparently published a few late tales in the Washington (Pa.) *Reporter*, lecturing and newspaper writing absorbed his attention after 1838. The publication of "Ashgrove, or The Return" in the November, 1838, issue of the *Casket* marked his final appearance in that journal and, presumably, the last of his border narratives. With new tasks at hand, Darby at last parted company with his alter ego, the frontier traveler Mark Bancroft.

# *Textual Note*

William Darby's magazine tales, most published under the pseudonym "Mark Bancroft," were never collected in a separate volume and only a few were ever reprinted in other periodicals. Yet these works tell us much about Darby himself, his view of frontier history, and his grasp of the literary conventions prevalent during his time. From a technical standpoint, even his best pieces display more crudeness than craftsmanship; but several manage to relate engaging stories, and a few combine effective dramatic action with an inventive use of historical fact. Among the latter are "The Hunter's Tale" and "The Moravian Indians," two readable tales that embody Darby's quintessential thematic concerns. Coincidentally both stories utilize a framing device involving a character named Kingsly (or Kingsley) Hale, a grizzled veteran and counterpart to Mark Bancroft.

As the author readily admitted, his prose lacked the refinement and elegance which was the mark of formal education. Hence, some minor, occasional emendations have been made in the interest of more intelligible texts. Darby's use of punctuation was erratic; in cases where the insertion or omission of punctuation has seemed likely to produce misreadings, commas or semicolons have been added, deleted, or substituted. Generally, however, Darby's punctuation has been allowed to stand, even where it creates awkward breaks in the sentence. His spelling was usually accurate, although he seems to have relished archaic forms of orthography. Obvious typographical errors have been silently corrected; but many misspellings and archaisms have been preserved, where the meaning is evident, for the sake of stylistic fidelity. Grammar apparently mystified Darby, and throughout his writing, one finds numerous fractured constructions. In cases where the meaning has been seriously obscured, corrections have been made, usually by the insertion of a single, bracketed word. Shifts in narrative perspective and confusing use of quotation marks pose the most severe problem for the reader. To avoid the ambiguity of quotes within quotes within quotes, I have eliminated one set of quotation marks and identified the narrator or speaker in brackets.

Finally, in Darby's defense, it must be said that rules of grammar, punctuation, and usage were not so strictly insisted upon in frontier America as they are in the modern academy. Like many contributors to the magazines of that era, Darby served no apprenticeship to literary art and saw his fiction as a casual diversion from other responsibilities.

# "The Hunter's Tale; or, Conrad Mayer and Susan Gray"

The text comes from the *Casket* of December, 1831. Like all of Darby's narratives, "The Hunter's Tale" is fraught with stylistic problems, but the story nevertheless contains a unified plot and some thrilling scenes. Here the author uses the familiar persona of Mark Bancroft, who plays the role of sympathetic listener as the veteran Kingsly Hale recounts the kernel narrative. The hero of Hale's account, Lewis Wetzel, actually performed the sort of feats described here and played a crucial role in the defense of the white settlements. Darby illustrates the awe which Wetzel inspired by depicting the effect of his name alone on the Indians and renegades. Few scenes in Darby's writing possess the horror of the massacre pictured in this tale; Conrad Mayer looks helplessly at the "still breathing and weltering forms of his parents" lying on the cabin floor. Darby inserted such detail not to produce sensation but to dramatize the real threat faced by pioneers. These grim events are balanced against the love story of Conrad and Susan, the sentimentality of which is entirely conventional—even to Darby's suggestion (borrowed from Chateaubriand and Byron) that theirs is a slightly incestuous passion.

Written for the Casket.

## THE HUNTER'S TALE; *OR, CONRAD MAYER AND SUSAN GRAY.*

*By chace our long-liv'd fathers earn'd their food;*
*Toil strung the nerves, and purify'd the blood;*
*But we, their sons, a pamper'd race of men,*
*Are dwindled down to threescore years and ten.*

Epistle to John Dryden.[1]

[Kingsly Hale]    Fifty years have flown, have flown away, since my infant feet traversed and since my infant eye ranged over the mountains, hills and vales of the Ohio regions, then emphatically called "THE WESTERN COUNTRY." It was amid those vast solitudes, that my young limbs were braced to climb the rocky steeps of the Monongahela woods, and there did my young eye first catch the beams of the morning, on the hills of Ohio. But if my steps were wild and errant as the deer I chased, my mind was led by the romance-like history of the place and time, to range over, dwell upon and strongly remember, scenes of human action still more wild than the then almost unbroken wilderness.

Fifty years are gone, and have borne with their seasons the Red men of the wilderness, and changed the wilderness itself to a garden. In this great change, where are the first race of whites who penetrated the wide waste, met and vanquished the native Indian, and dissipated the dark gloom? Gone to their rest, with a few remnants, of which I am one. The smile that lights my eye on seeing the thousand fountains of Ohio, as they are now to be seen, and when I mentally form the contrast with the past, is quenched by the tear of bitter remembrance.

The beloved friends, the protectors and companions of my infancy, where are they? With but a very slight change, I might repeat and apply to myself, the plaintive reflections of a man, who saw from a throne,

---

1. The epigraph comes from John Dryden's poem, "To my honor'd kinsman, John Driden." The quote is generally correct, though Darby spells "chace" and "purify'd" in his own rustic manner.

and wept over the evanescence of our best joys, our affections.

The friends of my infancy, where are they?— Where are the dear parents, authors of my existence? My brothers! they are no more; and thee, my tender sister, thou exist only in this sad heart— But! what do I say, where are entire families?— Cut down by the scythe of death.

In the extent of bereavement, I can mourn over a more, a much greater loss. My parents, my brothers and sisters were once twelve—where are they? one sister is left, the others are at rest, and their remains lie in the bosom of the west. There is a balm, though that balm may be moistened with the tear of regret, in recalling scenes long gone by, and in speaking of those forever loved but whom we can see on earth no more.

Amongst these friends whose eyes are sealed, the memory of none other returns with more warmth, than the rough warriors Conrad Mayer and Lewis Wetzel—

[Mark Bancroft]

The name of Lewis Wetzel, struck not alone my ear, for well did I also once know, the brown, gallant, brave, and generous hunter-warrior, and I started to my feet and seized the hand of the grey-haired soldier, Kingsly Hale, who was thus opening one of his "thousand and two" tales to a group of the most attentive young persons. Kingsly was a veteran who had seen much, suffered much; yet with the weight of sixty-five years on his head, his memory was little impaired, and his eye and tone of voice were still strong and expressive. My enthusiasm, though a stranger, was a spark thrown into a powder magazine, [and] it struck fire from the soul of Kingsly, who returning the pressure of my hand with more than my own ardor, exclaimed, "were you in this country when Lewis Wetzel and Conrad Mayer bore the rifle to the battle field?"

"Like yourself," I replied, "it is fifty years since I first set foot where we now stand."

The old veteran regarded me fixedly, the big tear trembled in his eye; but a moment restored him to himself, as he slowly repeated, as if to his own recollections,

"Fifty years! dreadful sounds—men and nations tremble at thy repetition—Stranger, for I cannot recall thy features, my name is Kingsly Hale;" "and mine, Mark Bancroft."

The recognition was instantaneous,—it was—might I say terribly pleasing? Forty years before had I seen and called Kingsly my friend, and what had we now to remember together? He regained composure first, when turning to the astonished group, some of whom were his grand children, and some were his nephews, resumed his tale, pointing at the same time down the placid Ohio, on the banks of which we were assembled; and to a point beyond the fine town of Wheeling which stretched along the landscape.

[Kingsly Hale]   My children, long before either of you saw the light, this now wrinkled Mark Bancroft and myself, sat under the shade on yonder hill and recounted much of what you are now to hear.

When in 1775, the Zane family built a fort amidst the plain on which that city now stands, for a city it is, in all the moral and social, and in every commercial attribute of a city. Wheeling Fort was the outpost of civilization.—The plain or bottom, narrow and darkened by trees and underwood, was over-shadowed by that hill, steep and impending also with a forest of poplar, oak, and other massy trunks, against whose columns the axe had never made its attacks. That creek now spanned by yonder bridge, wound its shaded stream behind that sharp and rocky ridge, gliding silently into the bosom of its mighty recipient, the Ohio. The great Ohio itself, the present channel of active life and commerce, was itself then an emblem of savage majesty. The stream was then, and perhaps in all former forgotten ages, as it is now, tranquil; but it was then solitary, and the view along its shores and current inspired feelings of sadness. Yonder western hills, beyond Wheeling Island, then rose bold, and blackened with an interminable forest. They were the eastern abutments of a boundless region, then with fearful import called "The Western Country;" or with still more awful import, "The Indian Country."

It was a country indeed, at the very aspect of which, the bravest heart felt a shudder; for, from its endless recesses, the ruthless and stealthy savage issued on his errand of death. It was a frontier, along which the Indian and white, the red and the pale warriors met, and often met in single and unwitnessed combat.

In the days of your grandfathers, we now sit on a spot they dared not visit without their terrible weapon, the rifle; nor did their rifle always save them from a foe who seemed to issue from the earth. But if the motion of the white hunter-warrior was slow, his march was steady and he sustained his post or fell; the white wave never flowed backwards towards its native ocean.

You have all heard of the Mayer and Wetzel families, for who on this side of the mountains has not heard of Conrad Mayer and Lewis Wetzel? But you may not all have heard how old Fred Mayer found his way to the banks of the Monongahela. Fred was a stubborn German, who, not liking the religion of his country, made one for himself, with a very short creed, and found it necessary to come to America to put his faith in practice. Fred brought with him some good share of Dutch scholarship, and a little gold, and what was far better than either, he brought with him a sweetly innocent and devoted wife. A few poor families came with Fred Mayer. They were peasants, stern, rough and muscular. Amid them, well do I remember the tear-eyed Maria Mayer; she was born to grace a court:—she became a flower of our wilderness. The little colony found a resting place on the banks of the Monongahela, and Fred and his Maria arrived just in time, for on the very next night afterwards was born their only son—their only child Conrad.

The morn which first dawned on Conrad, was a fine October Sabbath. Their church was the Monongahela woods, in which the new born boy received his baptismal name, and from which, thankful orisons rose to heaven for their safe arrival. Hardships had met them on their way, but sickness and death they had escaped, and now a son was born to share their future hopes.

We need not follow the infancy and youth of Conrad.

In despite of his father's attempt to teach him high German learning, this first born son of Fairstone rose to manhood, the active and untiring hunter, and the intrepid warrior. Thus he rose, or was rising, when the revolutionary war burst in distant and lengthened blasts, resounding from hamlet to hamlet, and from town to town, until its echoes were heard in the dales of the far distant west.—There was little need of repeated shouts of war to rouse young Conrad. From his father, he inherited a frame light and airy, but most powerfully strong and active. His soft blue eye bespoke the German, though his appearance and motions were French. His natural temper was wild and irascible, but his heart was tender. If he excited a tear from the eye of his mother, or of his foster sister Susan Gray, his kindness soon wiped that tear and remembrance away.

That heart must have been steel indeed who could have withstood the tears of either Maria Mayer, or her beautiful orphan foster child, Susan Gray. Very different hearts from steel animated the bosoms of Fred Mayer, and his son Conrad, and they were a family of love.

Susan Gray was the child of love and sorrow. Her father, Thomas Gray, the son of an opulent family near ———, married a lovely but poor girl, and indignant at the taunts of his family, sought the wilds of the west. The parents were unequal to meet the hardships of their new situation; they fell early victims, and the yet hardly lisping Susan, became the child of Fred and Maria and the sister of Conrad. The orphan shared the all of her protectors, and was vexed, and loved by the untoward but generous Conrad, who maintained at every shooting match that he had the prettiest sister in all America, and heaven-protected must needs have been the man who would have dared a contradiction; and another claim he had at the shooting match, of being the best shot over all Ten Mile and Wheeling woods; excepting, as some dared to whisper, Lewis Wetzel.

"Would I not give all my hunt this fall if I could ever meet this Lewis Wetzel"—grumbled Conrad, at a Redstone shooting match, as he overheard some one in a smothered voice say, "I wish Lewis Wetzel was here." Conrad bore away every prize, and swore he would "never shoot against another man until he met and beat the famed Wetzel."

The forest, hills, dales, and rocks, with the shooting matches, were the fields of fame of Conrad, from his boyhood, and before he had reached fifteen, he began to complain that bears and deers were becoming scarce; and at about sixteen his father removed to a valley on the head of Wheeling, near Ryerson's station. Accompanied by his faithful dog, several nights would sometimes intervene whilst this daring young prowler would sleep in the untenanted woods. His mother and Susan had always much chiding in reserve, which they always forgot between the return of his dog and himself, for Brawler always arrived first to announce his master.

On preparing for one of those expeditions, Conrad seemed to linger more than usual. He was uncommonly long in preparing his rifle and other accoutrements. He laughed, teazed Susan, and vexed his mother; but, as he often told me, an anxiety hung over him, he dreaded to leave home. The whole family shared the feeling and knew not why. The habitations were few, and far separated from each other; but as Indian war had not for many years reached those dells, no apparent danger seemed to impend, and yet the steady, firm, and everything-but-superstitious mind of Fred Mayer shrunk with dread. Fred Mayer had been many years a soldier, and felt ashamed of his own fears, laughed at himself and Conrad; and Conrad himself forced a playful catch, kissed his mother and Susan, and darted off for the woods.

The lingering form was not yet lost, for Conrad once or twice paused and looked back upon the paternal cottage, when his mother saw the ramrod of his rifle lying on their breakfast table. She seized the rod with an exclamation—she had time for no more—the rod and the light footed Susan were gone on the footsteps of Conrad.

The young hunter had disappeared from the cottage, and being at variance with his own thoughts, now hurried in the opposite extreme, and extended his pace to almost a run. His speed was soon checked as he heard his name anxiously pronounced, and turning, saw the airy form of Susan.

"You are a fine hunter," exclaimed the panting girl, holding up the rod. Conrad lowered his rifle hastily, saw his remissness, and forcing a gaiety he felt not, and patting the flushed cheek of the messenger, replied, "Poh! Susan, may be I left the ramrod behind to see if my sister would think [it] worth while to follow me with it."

"Conrad," rather solemnly, replied Susan, "do not call your poor little sister a fool—but—but come home with me; do not go hunting to-day."

"Ha! ha! Sukey, go home because I forgot my ramrod, ha! ha!"

"Conrad, I never saw you linger and turn back before," and the starting tear stood in her timid eye.

This appeal was always effectual in finding the heart of the otherwise wayward hunter, and setting his rifle against a tree, he seized the almost fainting girl in his arms, exclaiming with the most pathetic tone—

"Susan, if you were indeed my sister, I ought to return; but my heart tells me you are a thousand sisters in one, and ought I not to fly to the farthest woods, for I am only to thee a brother."

It was the moment they had found that there was a feeling between them infinitely more awakening, more anxious for each other, than that of brother and sister; but their looks spoke what their words dared not.

Susan gently extricating herself, exclaiming,— "My, only brother; if I stand here listening to such——I—I believe I must be"—and away she tripped, with sensations in which delight of heart greatly prevailed; and thus tripped to the summit of a small eminence, when turning round, she saw Conrad standing where she left him, intently gazing after her. They waved a farewell and parted.

All that day did Conrad, with steady steps and anxious feelings wind his way towards the Ohio. As the departing rays of day were leaving the earth in gloom, he reached the place where in two days his father had appointed to meet him with horses.

It was late in autumn; the morning was clear and bracing, and the limbs of Conrad, invigorated by rest on new fallen leaves, sallied forth, his rifle well poised on his shoulder, and Brawler, well trained, marching behind his feet, with the watchful eye and wary tread of a tyger. — Thus prepared, Conrad was treading slowly along the mountain-like hills, when spying a deer at a distance, he advanced with hunter caution, until within reach. The piece was pointed, and the unerring ball sped through the heart of the animal. But at the very moment when Conrad discharged his rifle, another prowler of the woods performed the same office, and the innocent buck fell by a double shot. Both hunter dogs preceded their masters, and commenced a furious battle over the prey; which was rapidly followed by a more serious contest between the two men.

The passions of Conrad, always excessively violent when roused, were raised to madness on seeing the stranger strike Brawler. "You cowardly villain, strike my dog, take that"—but active and athletic as he was, Conrad soon found himself engaged with an antagonist, who maintaining the utmost coolness, and also [being] a powerful man, demanded every exertion. For perhaps a minute the contest was doubtful, and entirely blinded by excessive rage, Conrad made repeated attempts to draw his knife. This, with the perhaps superior strength, and the perfect presence of mind he preserved, decided the contest in favour of the stranger, who at length by a skilful muscular exertion, laid the frothing Conrad prostrate, wrenched his knife from its scabbard and threw it to some distance, and then securing his arms, sat triumphant on his body. Pausing a moment for breath, and with the most provoking coolness viewing his still writhing enemy, [he] very calmly observed—

"Young man, whoever you are, your jerks can do yourself as little use, as me harm, nor do I intend to do you harm."

"Do me harm!" vociferated the prostrate hunter, in

accents of as much wrathful defiance as his exhausted frame would admit. "Let me up and on my feet; give me a chance, and we'll see who is to be harmed."

"As matters have thus far went," replied the collected and even smiling stranger, "I am accountable for my own acts; but as I have found you in a scrape and have never injured; why, I'll try to get you off safe."

"Insulting scoundrel let me up"—

"Wait, my good boy, until your fever cools."

"Villain," roared the now absolutely infuriated Conrad. "You dare not take your rifle and give me a fair shot."

"The poor young man is raving; bleeding will cool his fever," deliberately drawing his own knife.

The flashing blade no sooner met the eye of Conrad, who expected to meet its edge, than his rage was calmed in a moment. His eyes changed from an expression of rage to that of stern and even contemptuous defiance.— Not a fibre of his frame trembled; on the contrary, he steadily eyeing the victorious stranger, observed, "Murderer you make yourself; but let me advise in my turn. I fear you not, but if you have a father and mother, can you return to them and leave the body of an only son in the woods?"

"I have parents and friends, also," replied the stranger; "I never intended to injure a hair of your head, and as I see you are coming to your senses, you may rise, if your promise is given to act correctly. You are no coward, and I am no murderer; I cannot accept your challenge."

"It would be cowardice to betray such manly confidence," observed Conrad, as himself and [his] competitor rose to their feet. The two quadrupeds were in the mean time lying panting from their own share of the fray.

Having eyed each other a moment, after both had resumed their arms, the stranger very good-naturedly observed—

"My good friend, we have made a very lucky escape, and have much reason to remember each other; and as I have as much reason as yourself to be ashamed of so rash an act, we may exchange forgiveness."

Conrad felt as many others have felt who have been

doubly vanquished, and with all his really strong feelings of generosity, a lurking mortification gave a sulky moroseness to his manner, as he rather ungraciously replied, "I suppose I am to be thankful for not having my throat cut."

"Have a care your fever does not return and affect your brain again, young man;" very slowly and provokingly replied the stranger.

This was too much for the chafed spirit of Conrad, who commenced reloading his rifle with violent gestures and feelings of anger. His opponent also, but with the utmost coolness commenced a similar operation, and long before the enraged Conrad had his weapon prepared, the stranger with a half suppressed smile, was very composedly eyeing the rash young hunter, whilst standing grasping in his left hand his well loaded and primed rifle, and patting the head of his wounded dog with his right. When Conrad had put his weapon in order, the stranger then observed:

"You say, friend, that you are an only son; I am a little inclined to think your father and mother would soon be childless if your life depended on which of us could load our rifles first. Be calm and hear me," continued the stranger, "before you attempt again to grapple with a stranger who has given you no good cause; permit me to give you a lesson. Do you see a white spot on that hickory tree yonder?"

"I am not blind," sulkily replied Conrad.

"Except with useless passion," replied the stranger, as he raised his piece to his face, and in a moment the white spot was gone, and the intrepid and manly hunter stood with his empty rifle smiling in the face of his now abashed companion, who remained an instant absorbed in silent wonder; at length [he] ejaculated with great warmth—

"Well! well! this is too much, I am conquered."

"But alive yet," replied the stranger, as he walked swiftly towards the tree into which his bullet was lodged, which having reached, he held up his rifle in his left hand, shouting, "you see she is empty."

"I do," replied Conrad.

"Now then come here," continued the stranger, "and fill up this bullet hole, and then stand on one side." Conrad silently obeyed the order, when the stranger drawing his tomahawk, made a blaze, in the centre of which he made a small black spot with powder, and then laughingly observed— "young man you will now see, what, may be, you never saw before"; placing his back to the marked sapling and grasping his rifle, with the muzzle forward in his left hand, [he] bounded from the tree with the speed of an elk.

The wonder-stricken Conrad stood immovable, until he was roused to exclaim in extreme astonishment, "who can he be?" on seeing the stranger suddenly stop, wheel and fire. The report of the rifle and disappearance of the mark, began to excite feelings of almost superstitious dread in the bosom of Conrad; feelings which were wound to their height as the terrible stranger returned, running with uncommon speed, and coming up, handed Conrad a completely loaded rifle.

Eyeing the rifle and the owner alternately, Conrad at length found breath to exclaim.

"If you had not the look of a fine young man, I should suppose"—

"I was something worse;" replied the stranger, "but it is time we knew each other."

"My name," with some hesitation, replied Conrad, "is Mayer. I am the—I am sorry to say, undutiful son Conrad, of Frederick and Maria Mayer."

"And I am not worth the name, perhaps," said the stranger, "but I am Lewis Wetzel."

The arms of Conrad were instantly round his preserver; for it was the wind-beaten and sun-embrowned hunter-warrior, Lewis Wetzel, with whom he had been contending.

Their mutual embarrassment having a little subsided, Lewis observed—

"Conrad, as you have found I am a man just like yourself, suppose we have our breakfast; we have earned it. Let us skin this chap, and carry his carcase to my camp. We have been playing the fool long enough to be hungry."

On the bank of a clear stream, the trees for a roof, the two hunters feasted; gave each to the other a short account of their lives, laughed and spent the day; for that day they did not resume the chace; and when evening closed upon them, they slept on their leafy couch as if nothing of consequence had passed between them; and while they sleep and hunt, let us wander up Wheeling and visit the cottage of Fred Mayer.

The two days after the departure of Conrad, were cool, and until towards the evening of the second, clear. For the next morning Fred had prepared every thing necessary to set out to meet his son. Towards sunset, the wind set in from the northeast; the whole heaven became overcast, and night set in raw and cold, and that most dismal of all domestic sounds, the howl of the house dog[,] mingled with the night blast. The family had, in some measure, conquered the sense of lonesomeness, which is so painful when a few human beings gaze upon each other for the first time, and feel that they are a defenceless few alone in a wilderness. Over the hilly and variegated peninsula, between the Monongahela and Ohio rivers, at the early day of our tale, the fields were small; they were few, and they were far distant from each other. The cabins were rude and often constructed as blockhouses, for defense. The almost imperceptible paths wound through interminable forests, where almost every sound which broke the silence, was of the appalling kind. It was these lone habitations which became so often the scenes of savage murder. This is not the product of imagination; it is the bitter remembrance of real life and death, the remembrance of the worst features of human strife. Fifty years have passed and snowed upon this head, yet it seems only yesterday, the dark and dreadful night, when Fred Mayer and his wife and child, [were] far removed from every other eye, but that Eye which never sleeps. The night passed slowly away, sleep they could not; each tried to convince the other, and say to their own hearts, "it is the absence of Conrad." But their champion had often been absent before; their heaviness of heart had now something of distressing beyond all former anxiety for their Conrad. Towards midnight the wind entirely ceased, rain began to patter on the roof, and the darkness, heavy before, became still more dense. The howl of the watch dog became more loud, and anxious in its tones. Thus passed the night until the faint grey light of morning began to dawn.

"God be praised," sighed Frederick, "it is break of day."

At that moment the faithful sentinel at the door, by a fierce and rapid barking, announced the approach of some living object. The warning voice was as rapidly followed by a scream, a few groans, and all again was silent.

Frederick Mayer, like all truly brave men, lost the sense of undefined fear at the aspect of real danger, sprung from the bed with intent to seize his rifle, in the use of which he was no bungler; was it accident, he did grasp the rifle, but his foot struck a log of wood and he fell to his knees as the thin clapboard door was dashed from its hinges, and three rifles discharged into the cabin in rapid succession. The most heart-rending screams roused Frederick to frenzy. The great muscular force of his youth seemed to be redoubled. The Indians were deceived by his fall, and naturally concluded their victim safe. They were soon undeceived as uttering the names of his wife and child in a voice of absolute fury, he rose to his feet and, firing into the group, attempted to turn the butt of his rifle. The shot took effect on one enemy, but the stock of the piece flew to shivers against a joist as the owner was grappled with and thrown on the floor. His presence of mind never for an instant forsook him, and feeling that though one against such fearful odds of numbers, that his enemies were exposed to the danger of wounding each other, which in effect took place. Firmly grasping his formidable weapon, the naked rifle barrel, and turning himself by main strength on his face, [he] once more regained his feet, and by a sweep of the iron bar carried away the entire upper part of the skull of another Indian.

The cabin was now become indeed a scene of indescribable horror. The whole events I have noticed did not occupy more, if as much, as half a minute. The screeching Susan was dragged by the hair at the very moment that her protector fell in the first instance. The maddening sight was the last that Fred Mayer got of any part of his family until the tragedy closed. The groans of his wounded wife he heard amid the combat, but he saw her not; the bed on which she lay had been broken down, and her pure blood mingled with that of her savage enemy. The very best safeguard of a single man against many was thrown round Fred; that is, he lost all sense of self-preservation, and bent the whole force of his body and resources of his mind on the destruction of the destroyers of his family—and how the contest would have terminated we can never know, as the shouts of other voices now mingled in the maddening fray.

You may remember, my young friends, we left Conrad and his new made friend, Lewis, sleeping on a rivulet of Ohio. Let us return to them and watch their motions.— Next morning after the scuffle and happy reconciliation, the sun shone clear upon the heads of the two children of the woods. Conrad attempted to make amends for the sallies of the day before—it was an effort understood by the keen-eyed Lewis.

"My dreams hang *heavy* on me this morning, Conrad," said his companion, "and with all your laughing your brow is *heavy*. Have you ever sought the trail of the Indian?"

"I have not," replied Conrad.

"Then walk backwards, and carefully put yourself into that tree top," pointing to a very large oak which had fallen with its leaves on the previous summer—"and remain there with your rifle prepared until I return."

Conrad eyed the speaker, but found an air of command which he felt he ought to obey, and he did obey. Lewis then left their camp with a tread that gave no noise from the early fallen leaf. His course was northwardly and towards Shepherd's Fort. Hour followed hour until after mid-day, as the impatient Conrad watched the return of his companion in the direction of his departure. He was intently looking at a waving bush on a distant hill, doubtful whether it was a man or not, when he felt his shoulder struck, and "Ingens are not Deer" came from Wetzel, who had thus given a lesson of vigilance.

"Prepare, Conrad, seven or eight of those black rascals are gone in the direction of your father's house."

"Good God!" exclaimed Conrad, "poor Susan, why did I not go home with you? My sister, my father and mother."

"Standing there making speeches will do no good to either your sister, your father or mother." Then pausing a moment and with his compressed mouth and expression of features which no man, however firm might be his nerves, ever beheld without feeling a something saying, "Let that man never be my enemy"—[he] muttered with appalling emphasis, "If the whole of these cutthroats ever again cross the Ohio.—why, they'll conclude that the D——l and Lewis Wetzel have had a quarrel lately; but I'll try to show them that myself and an old friend are not separated yet."

Little more was said; a few slices of half roasted venison were cut from the residue of the deer slain the day before, and the two hunters were with careful but rather rapid steps measuring their way to the northeastward, with a view, as Lewis whispered, "to fall in the rear of the *Ingens*." With all their untiring speed it was evening when, reaching the head of a hollow overspread with the rank growth of the past summer, that Lewis stopt suddenly, and pointing with his ram-rod to marks Conrad could scarcely perceive, observed in an under tone, "Here, here! are their trail!" They had been several hours far within the range where every stream and ridge was known to Conrad, whose inward agony of mind increased at every moment, in following the steps of his wary leader, and [he] saw him advancing in the direct course towards the home of his parents and sister.— He was almost provoked at the cool and undisturbed behaviour of Lewis, but the dreadful appearances made him completely submissive to the orders which were given with a confidence which inspired hope in the very face of despair.

I have already told you, my children, that the evening was heavy, and the night unusually dark. That darkness closed upon the—I might say, angels of deliverance, some miles short of Mayer's cabin. It was on the closing of light that any expression of impatience was shown by Lewis.

"Must these villains escape?" The expression was lofty, and calculated to alarm Conrad; but the long tried warrior repaired his mistake with admirable quickness by adding, "till to-morrow morning," whispering at the same time, that "It is not the custom of the *Ingens* to attack only at break of day."

Still they advanced slow, silent, and listening at every few steps. For some hours the wind enabled Lewis to keep his course, but when that guide failed, and the black and covered sky hid every star, the bark of the trees were felt.

"In any common case," again whispered Lewis, "maybe I could find my way, we must be near your father's and we may pass it, we must stop."

A sigh and shudder was all the answer Conrad could make, and they crouched down beside two trees. I need not say hours were weeks, as both their faces were turned to what they thought the east. It was an opening amongst the trees, which at last began to widen, the trunks and large branches began to appear. Lewis was just ready to say, "Break of day," but was prevented by Conrad springing to his feet exclaiming, "By heavens!" His loud expression was promptly and effectually arrested by the powerful hand of Wetzel, who almost jerked him off his feet.

Conrad, brought to himself, in a hurried but suppressed tone informed Lewis, that they were between two and three miles from his father's house, that the opening they saw was an abandoned settlement. They were on their way before Conrad had finished. Avoiding the open old field, they were soon on a cattle path, and in a few minutes, on rising a hill, the long drawn howl of the house dog was on the point of being answered by the two brute sharers of their march, but a touch of the ramrod reduced to complete silence the well-trained mastiffs. Their speed was every moment increased as the cry of the watchdog became more and more distinct.

Suddenly Lewis stopt, and, listening a second or two to the change of note of the dog, then most earnestly observed to Conrad—

"Now, my brave young man, follow my directions— Your house is surrounded by these savages; advance cautiously and do not fire unless sure of your mark. When you do fire, instantly retreat and reload; but of all things do not for any cause rush towards the house unless you see me."

The orders were cut short; the death-scream of the dog, the equally terrible silence which followed, and then the rapid firing, and the screams of the females, put all farther delay out of question, and yet the never-disturbed peace of mind of Wetzel, as Conrad afterwards told me, had more the appearance of a man advancing on a wounded bear than on he knew not how many armed men. It was at the moment when a blow from a tomahawk sunk the brave old Mayer, that the voice of Lewis Wetzel was no longer heard in whispers, but echoed to the surrounding forest. "Conrad, your family is murdered. Revenge! revenge!" and shouted his own name with a force almost beyond human. If an earthquake had burst beneath their feet, the effect would not have been more terrific on the minds of the *Ingens*, as he called them. They who yet survived rushed from the cabin, at the threshold of which two fell to their sleep of death, and the astonished Lewis saw only one flying savage. "You shall follow," as he gritted his teeth in rage, and darted after his, to him, certain prey. For once even the consummate skill of Wetzel was within a hair breadth of failing. The Indian's piece had not been discharged, and knowing that both white men had discharged their rifles, and finding himself pursued by only a single man, who was every step gaining upon him, the savage sprang to a tree. Lewis saw his error, and as the piece was raised he fell prostrate, at the instant the ball passed through his hunting shirt above his shoulder.

The Indian was now in his power, but without discharging his piece he grasped it in his left hand, and in a few hundred yards the Indian was a corpse under his hatchet.

The sun had not yet risen when Lewis returned with wary steps towards the cabin. To the name of "Conrad," called in a voice louder and louder, no answer was given, and he finally reached the dreadful spot stained with the blood of six human beings. With his back to the fireplace stood Conrad, his eyes fixed in horror on the still breathing and weltering forms of his parents. To the friendly and now touching voice of Lewis no answer was given, and even Lewis himself, accustomed as he was to the dread horrors of savage war, could not avoid exclaiming, "Is all this real."

"Yes, real," replied Conrad, with a bursting sigh, "and my fault." A vast passionate flood of tears followed, but that flood was salutary. Conrad was restored to himself, if a man inflamed to almost the madness of rage could be said to be restored. "To the woods I fly with you, Lewis, the Indians' blood shall pay for this— but oh! Lewis—can I ask"—

"For the body of your sister," interrupted Lewis, "she is not dead, but a prisoner, in my opinion. Your horses are gone, for in returning to the house I had the caution to examine the stable, where the tracks of men and horses are plenty. It is all strange—very strange. There were more men on this murdering party than we have found. It is strange—very strange."

"They may be lurking near," replied Conrad.

"They are making their way to the Ohio," bitterly interrupted Lewis, "if I did not know these wolves I would not stand here."  *   *   *   *

[Mark Bancroft]

Here Kingsley paused as his young auditory awaited the finishing of his story.

"I am talking about events in a different age from the present," at length he resumed.

[Kingsly Hale] Before the parley I have related, short as it was, was closed, Conrad Mayer had no living parent.

"I am alone! I am alone! Susan, my Susan, I follow thee."

"And I am with thee to the Shawnee towns," replied Wetzel, who commenced to place the dead bodies of Mayer and his wife side by side, covering them with the bed clothes, and after swallowing a few hasty morsels the two persevering warriors were again on their way in pursuit.

Lewis traced the horse tracks, which for several miles were found along a path towards where Waynesburg now stands, and then bent to the southwestward over the southern heads of Wheeling into the valley of Fish Creek. — The tracks proved haste and the small puddles left where water courses were passed enabled Lewis to determine, as he vehemently expressed himself, that—

"These painted scoundrels are gaining from us."

In our days, when our *fine* young men must *ride* along good roads, and would shrink at a walk from Wheeling to Washington in Pennsylvania, you may well feel astonished when I tell you that with all the fatigue of the day and night before, Conrad Mayer and Lewis Wetzel were again on the Ohio before night closed on their path; but they arrived only in time to find their objects of pursuit had crossed that great stream.

Arrived on the bank, Lewis, turning to his companion, observed—

"Conrad, we must sleep, if we do sleep, on yonder bank."

"I'll be on that bank this night if I swim," replied Conrad.

"And swim you must, but we must take care of our arms. Conrad, had you or any of your family ever a worthless, cowardly enemy who fled from you to the Indians?"

The question was a volume at once to Conrad, who, clapping his hand to his forehead, reflected in silence for several minutes, and at length answered—

"Yes, there was one fellow, Ned Trash, from whom I

won a hunt of deer skins at a shooting match, and afterwards knocked down for saying he intended to court
Susan. He has been gone upwards of two years."

"And is the worst Shawnee in the towns, and has got
your Susan without courting. The moment I saw the
stable this morning I set down in my mind that the cowards who left their companions were not *Ingens*. Look at
yon fire." And a fire was now distinctly seen amongst
the trees on the opposite shore.

"Blood painted monsters," muttered Lewis, "you left
a home, flowing with blood this morning, and to-
morrow morning your blood shall flow. My friend,
Conrad, we'll cross the river, and do you take care of
your rifle and your girl, your sister, or what you choose
to call her, and I'll lead those new made *Ingens* a dance—
never mind if I dont."

Though Conrad felt very much disposed to lead them
a dance himself, his increasing confidence in his commander kept him silent and submissive. The river was
passed, and as Lewis intended, they made land far
enough below the savage camp to secure themselves
from discovery. Short as was the distance, however, it
was far in the night before the dying embers of their fire
and the sleeping bodies of the enemy were seen by the
two cautious hunters, who in their approach kept a
deathlike silence. There was indeed but one sound of
human voice which broke upon the dreary scene. That
sound was the heart-broken and despairing aspirations
of the captive girl. Though bred in a forest Susan Gray
was reared tenderly. The first rude shock that marked
her young days from the death of her parents, was one
of utter destruction. The gray dawn of morning was the
messenger of horror. The faint light broke over the eastern hills of Ohio,—broke over those native hills she
dared not hope ever again to behold.

The greatest danger and no danger produce the same
effects, says Zimmerman,[2] quoting Count Lippe. Many

---

2. Darby here refers to an epigram by the Swiss historian, philosopher, and aphorist
Johann Georg Zimmerman (1728–1795), whose writings on Prussian history probably furnished the source for this comment.

are the instances I have known where that truth was shown by Indian captives, and more than one when it proved their last defence.

With the dawn the savages arose; one sat gloomy and with a visage of more than Indian ferocity opposite to where the slightly bound captive was lying. Her eye caught the glance of the villain, and the effect was an exclamation of indignant contempt rather than that of what might in common cases be expected.

"You are Edward Trash[3]—where are my father and mother?"

Though steeped in blood, Trash cowered under the glance of an angel and the voice of God, for what was she at this awful moment but the messenger of him whose eye pierces the thickest darkness.

"Yes!"—she reiterated. But her voice was lost in sounds equally the voice of Heaven. Trash, like all cowards who shrink from the moral power, [but] soon gain confidence when assured of the personal weakness of their opponents, sprang to his feet with an expression of rage and hatred which only a renegado ever can assume. What would have been his next act we can now never know. The whole scene was rapid as the flash and report of an impending thunder cloud, and followed by the report of a rifle and the dreadful name of Lewis Wetzel resounding along the Ohio.

Trash fell under the ball of Conrad, who with burning impatience awaited the signal to fire. It may seem strange that Lewis did not also fire, but he was too good a warrior to give his enemies an advantage. His name, terrible to the real Indians, was tenfold more so to the Girtys and other whites, who, though never equal to the Indians in their mode of war, proved that the Shawnees could be far outdone in brutal cruelty. Urged as they were to flight by the avenger, the two unwounded cap-

3. Edward "Ned" Trash seems clearly based on the renegade Simon Girty, to whom Darby alludes two paragraphs later. Girty, also mentioned in "The Moravian Indians," was one of the authentic villains of border history. A renegade adopted by the Senecas, he took delight in torturing white captives; Richard Slotkin has recently described Girty as the "antitype" of the hunter-warrior, embodying "all the negative, evil possibilities inherent in emigration to the wilderness." See *Regeneration Through Violence: The Mythology of the American Frontier, 1600–1860* (Middletown, Conn., 1973), 291.

tors of Susan gave to their pursuers an advantage, which on another occasion as well as on the present, was most effectually used by Lewis Wetzel. The two fugitives separated, or more distinctly, one outran the other, and left each to contend singlehanded with a man, that, perhaps, if in the woods of Ohio, and armed only with rifles, there was not then on the face of the earth, another who could have contended with the least probability of success. The wary Lewis in the outset did not exert his utmost powers of speed, but awaited the very effect I have noticed, but that effect once produced, every muscle was strained, and every moment the hindmost savage heard, or thought he heard, nearer and nearer the rapid tread of his pursuer.

Amongst the unequalled combination of qualities as a warrior, possessed by Lewis Wetzel, one of the most remarkable was the skill with which he drew his enemy's fire—an advantage he scarce ever failed to obtain, and almost certain death was the consequence to his opponent, for an empty rifle in his hand was most dangerous to those thrown off their guard, and his ability of reloading in full retreat or advance, enabled him to deceive his adversary by actually throwing away his own fire, a stratagem he put in practice in the present pursuit. Before he could secure an unerring aim he halted and discharged his piece. The discharge and wheeling of the nearest savage was the work of a moment. Affecting to retreat in turn, the triumphant shouts of one enemy and sound of the rifle deceived the other, and both rushed on to expected victory—but certain destruction. The moments were few until Lewis was again ready to "*take tree,*" but as he sprung behind one, the foremost *Ingen* contemptuously shouted "Ingen not a fool."

"A doe skin would be too high a price for your wisdom," muttered Lewis, his eyes flashing like a tyger's, as he awaited the approaching monster.

"Empty gun, white man."

"Empty through your heart, murderer, once white man," passed through the gritting teeth of Lewis, and one frightful groan, the last of the renegado, seemed an echo to the sharp crack of the rifle.

The remaining monster in human form was now too far advanced to retreat with any safety, and, rendered desperate, it became now a real struggle for life and death, and had the enemy been truly an Indian, even the skill and activity of Wetzel might have failed. Both feeling that their blood depended on the issue, [they] put every nerve and sinew to the strain. The disadvantage was fearfully on the side of Wetzel, and must have been fatal had his pursuer not been determined to make security more secure, [and] reserved his fire awaiting a chance of discharging with certain aim. Thus proceeded the race for a few hundred yards, when Lewis once more "*tree'd.*"

"You're a dead man," roared a thundering voice.

"H—l shall have one tenant more," seemed to come hoarsely from the bowels of the earth, as Lewis lay like a couching lion. Not conceiving the possibility of encountering a loaded rifle discharged not three minutes before, with dreadful oaths expressed in *good* or *bad* English as you choose, his adversary advanced.

"I'll know who this scoundrel is before I finish him," muttered Lewis, as he deliberately sent a ball through his left arm and shoulder, and dropping his rifle, seized his tomahawk and rushed upon the fallen. "Never carelessly approach a wounded enemy," was a maxim Lewis had good reason to remember, as he learned its wisdom from seeing a rifle muzzle raised and pointed to his breast. The rifle ball and his tomahawk passed each other in mid air. The ball passed harmless, but the hatchet lodged in the brain, and forever concealed the last of the captors of Susan Gray.

[Mark Bancroft]

Here Kingsley stopped as if his tale was ended. Every hearer felt burning to know what became of Conrad and Susan, and at length finding the old historian silent, more than one voice rather impatiently breathed, "And Conrad and Susan[?]"

"Reared a fine family of young Conrads and Susans," resumed Kingsley. They returned to see the fresh graves of their parents which had been laid in earth by a party of men the very day of their return. It was long before their hearts could again feel the gaiety of former days, but time softened the memory of the past, and when the name of their eldest son, Wetzel Mayer, was pronounced, they remembered the warrior of the west, and many is the time that young Mayer ran to meet the coming warrior, and many is the time that Susan Mayer sent to heaven the breathings of innocence mingled with like aspirations from other mothers, for the preservation of the brown warrior sleeping in the far distant woods of Ohio or Muskingum.

The reader might view some of the incidents of the preceding tale as so far bordering on the marvellous as to be out of nature:— But there is not one incident but in some case or other really happened within forty miles of Wheeling between 1770 and 1795. The power of loading a rifle whilst running in woods was really possessed as represented in the person of Lewis Wetzel, and actually exercised not very materially different from the incidents related in the tale. My object has been to represent a Hunter Warrior as they were in fact, without putting slang and vulgar *patois* in his mouth, never used by him. I was bred among the hunter warriors, and have seen and heard them in all situations, except that of war. I have seen them serious and sad, and have seen them in their hours of revelry, and when shouldering their rifles for chace and war.

MARK BANCROFT.

# "The Moravian Indians"

In a letter to Lyman C. Draper (July 24, 1846), Darby noted that "The Moravian Indians" had not been conceived as a tale: "That subject I conscientiously treated as History." Two events—the massacre of the Moravian Indians in March, 1782, by a Washington County militia under Colonel David Williamson and the subsequent disastrous expedition of Colonel William Crawford—indeed form the historical context of this narrative. Yet Darby depicts these events in relation to the fictional story of Saul Garvin, a white man who dies with his Moravian brothers. The narrative opens with Kingsley Hale's recollection of the thwarted romance which in 1776 prompted Garvin to forsake the white settlements; the latter portion unfolds in 1799, when Hale learns of Garvin's fate from an old Moravian chief who has returned to the site of the massacre. Darby believed that the events surrounding the slaughter had been "discolored," and here he laid the groundwork for a journalistic campaign to exculpate Williamson (a friend of the Darby family) by glossing over the grisly event and thus minimizing the commander's responsibility. The author also sought to correct the account promulgated by the Moravian missionary, the Reverend Heckewelder, who (in Darby's view) failed to see the incident in the context of ongoing, mutual atrocities. To dramatize this point, Darby carried forward the action to include the torture and execution of Colonel Crawford. The text comes from the *Casket* for May, 1833.

Written for the Casket.

## Preparatory Remarks to the MORAVIAN INDIANS: A TALE.

The murder of the Moravian, or Christian Indians on the Tuscarawas, in 1782, was amongst those acts which make a nation blush; but like all other acts of man it has been discolored. The name of Col. David Williamson, who was the nominal commander of the party who were the perpetrators, has been held up to infamy as a monster. This preface, and the Tale which follows, were neither of them written to excuse the deed of horror, nor have I ever heard a single voice raised in its justification, though I was bred from a child to mature years near Washington, Pennsylvania, and of course in the very section of country from whence the actors proceeded.

The Christian Indians were placed in the very most dangerous position that was possible, not, as commonly supposed, on the Muskingum, but Tuscarawas, directly between the warlike tribes and the equally warlike frontiers of Virginia and Pennsylvania; and as a natural and inevitable consequence, exposed to the suspicions of both parties.

In the work published in 1819, in Philadelphia, by the A. P. S., and written by Mr. Heckewelder, formerly a Moravian Missionary among the Indians, we are made to believe, as far as the context can influence our opinions, that the Christian Indians on the Tuscarawas, were safe except on the part of the whites.[1] This was far indeed from being the true state in which these people were placed. The Simon Girty mentioned by Mr. Heckewelder, was then a renegado amongst the hostile Mingoes and Shawnees, and in deeds of blood suffered no man to be his superior.

The almost universal opinion on the frontiers of Virginia and Pennsylvania, from about 1778, was, that the chieftain and peaceable Indians or Tuscarawas ought to be removed. The lawless bands on both sides were dreaded, and the considerate and humane part of the whites, an immense majority of the whole, in foreseeing, most anxiously desired to avert a catastrophe.— With many of the actors I was personally acquainted, and

---

1. As these prefatory remarks suggest, Darby found John Heckewelder's version of border history somewhat slanted. He quibbles with details from Heckewelder's *Account of the History, Manners, and Customs of the Indian Nations* (Philadelphia, 1819).

must say, that the result of the expedition could never have been premeditated, except by a few if by any single person. The act was loudly, and I may say almost universally condemned in the settlements, not simply from dread of revenge, but from genuine feelings of humanity.

Beside giving a coloring too strong on one side, Mr. Heckewelder has made some material errors in facts and dates. In page sixty-four this author quotes part of a speech made by an Indian, at which he says he was present, April, 1787; and in the next page states that "Eleven months after this speech was delivered, ninety-six of the same Christian Indians, about sixty of them women and children, were murdered, &c." It may be rationally conjectured, that the two last figures ought to be transposed, and make the date 1778; but even then the date of the massacre would fall in March, 1779, whilst it really took place in the summer of 1782. Without troubling the reader with personal detail, I can assert that, though very young, I cannot be mistaken in the latter date.

In page two hundred and eighty, when speaking of the second party sent out the same summer, 1782, but commanded by Crawford and Williamson, Mr. Heckewelder states, after giving some previous movements, they then shaped their course towards the hostile Indian villages, where being, contrary to their expectations, furiously attacked, Williamson and his band *took the advantage of a dark night and ran off, and the whole party escaped* except one Colonel Crawford and another, who being taken by the Indians, were carried in triumph to their villages, &c.— This account is exceedingly incorrect. There were several other persons taken with Colonel Crawford and tortured to death. Two very remarkable escapes were made; one by a man of the name of Stover, and the other, Dr. Knight.—The adventures of these two men would figure in romance, with all the interest of truth. Stover was well acquainted with the country, and reached the Ohio on the fifth day. Dr. Knight was about twenty-two or twenty-three days exposed to every hardship and danger.

So far again was the party from escaping with the exceptions given above, to say nothing of others, there were three men out of the near neighborhood where I lived who were never again heard of—of course perished; their names were William Nimmons, William Johnson, and William Houston.

These historical facts are given to serve as data to explain the natural causes of a deplorable event. In the Tale my object has been to paint the times, and give the feelings of men as they were then agitated. Those feelings had their play in the presence at an age when impressions are not simply deep, but indelible. I can at this moment, when upwards of fifty years have passed, see the faces of my mother and another woman who came running to where my mother stood, crying— "James Workman is killed! Oh, James Workman is killed!" Mr. Workman was not, however, killed—he returned to his family.

## THE MORAVIAN INDIANS

> *I cannot weep, yet I can feel*
> *The pangs that rend a parent's breast:*
> *But ah, what sighs or tears can heal*
> *Thy griefs, and wake the slumberer's rest?*
>
> McDiarmid.[2]

[Mark Bancroft]

If ever the view of any one picture of human improvement was more than all others, calculated to inspire sentiments of the most sublime enthusiasm, it is that of the "Great West," that immense region around us, and from which issue the thousand and ten thousand fountains, mingling their tribute to form the mighty Mississippi; but to feel the entire beauties of this canvas in their full harmony, they might have been seen as I have seen, when the first outlines of civilization were sketched; and now, when are presented in continual succession, farms, towns and cities, connected by rivers, roads and canals, with all the busy hum of commercial life. They must have been seen whilst the howl of the savage was still heard in the dark wild—when barbarism is replaced by all the allurements of cultivated society. "Yes; at the extremes of sixty years have I traversed these banks," said old Kingsley Hale, raising his voice, while his still expressive eye glanced down the tranquil Ohio, and with his hand stretched towards the rising city of Wheeling.[3]

"My young friends, let any one of you imagine himself encamped on this spot alone and with an unbroken forest around him, crouched under a fallen tree with his rifle, his only friend, clasped to his breast. It is night, stillness and darkness reign over the waste. You are

2. The epigraph comes from the opening lines of John McDiarmid's poem "To the Memory of a Very Promising Child," published in McDiarmid's popular miscellany, *The Scrap Book* (Edinburgh, 1822), 451.

3. Kingsley (or Kingsly) Hale also recounts "The Hunter's Tale." In contrast to Mark Bancroft, the frontier traveler, Hale is a long-time resident of the Wheeling area and a veteran of the border wars.

fallen into a slumber, from which you are aroused by a sound, long and piercing; it comes from beyond that river. Is it the Cougar's scream? no. It is a thousand times more terrible—it is the yell of an Indian. This sound ceases, stillness again reigns; fatigue wraps your senses in sleep, from which the burning rays of a summer sun recall you to waking recollection; you grasp your faithful rifle and rise with caution; you dare not stir a leaf, but what do you see? One wide sweep of cultivation. The forest is broken, fields stretch beyond fields, and of the primeval woods, what remain, serves to form a part only of the enchanting landscape. A city, with all the attributes of wealth and human enjoyment, occupies the foreground. What would be your astonishment at such a change? You could not believe it other than illusion, for such a change have I seen, and from this very spot. Along this bank I was one of four, the remnant of forty, who escaped the savages. On yonder bank, and under that immense hotel, slumber the dust of my fallen friends."

"Grand-papa," interrupted a lovely girl, "you have put us all into such a melancholy kind of—joy; such a—oh! I don't know what to call it; but you promised us the tale of Schoenbrun."[4]

"My little Ellen," said the old warrior, placing his hand on her head, "the tale of Schoenbrun will indeed give melancholy joy—it will excite regret for the past, and gladness that those days of blood are long gone into years of past time.

[Kingsley Hale]    In Europe, my young friends—my children, amongst many other societies of christians, arose one, "the United Brothers." Some of these men came to America, not to bring a sword, but the glad tidings of peace. Their persuasive voice reached into the deepest recesses of these woods of which I have spoken, and entered the hearts of many natives, who embraced, not in name, but reality, the doctrines of Christ. Of these red men and their families, many settled on the Muskingum. The messengers

4. The actual site of the massacre was Gnadenhutten—a fact Darby clearly knew, for he mentioned the episode also in "Reminiscences of the West."

of christianity were Germans, and in memory of their native places, German names were given to three villages, Salem, Gnadenhutten, and Schoenbrun. During the twelve or thirteen years which passed between the "Old French War," to the beginning of that of the Revolution, peace reigned over these solitary settlements; they were spots, and pleasing ones, on the beautiful Tuscarawas, where the children of nature learned to lisp the name of Him whose power brought them into existence. They were spots on which the eye of benevolence delighted to dwell, but over which the prophetic eye would have wept tears of bitterness. In many of my hunting excursions (for then we were all hunters), I strayed to the creeks of Tuscarawas; and many is the night my weary limbs found rest at Schoenbrun. But those days of peace were to be succeeded by a storm— a sweeping destruction, the American Revolutionary War. That great period gave a republic to the earth, and humbled the proud oppressor. Such benefits were purchased with blood, and not in every case sustained with blood.

If a few native Indians planted the olive, the much greater number cherished the laurel, and remained ready to dig up the tomahawk, or hatchet of war. With an improvidence, which has cost our lengthened frontier so much of blood and misery, these warriors were left to our enemy. Every art was used to excite them against us.

Slow and constant was the stream of white emigration; and with superior arms and other means, every rising young man became a natural enemy to the Indians, and the Indians felt that their inheritance was passing to the white race. Between these warlike bodies stood the three defenceless villages of Gnadenhutten, Schoenbrun and Salem. Threats from the east, at first slight, but yearly becoming more fierce and loud, reached the Moravian Indians on Tuscarawas, in accents of death. The sounds from the west were not less appalling; the Christian Indians stood between two hostile nations, suspected by, and exposed to the vengeance of both— no government to offer an arm to these unprotected and unsuspecting people; and the whirlwind of destruction reached their dwellings. They perished, but not alone.

Let us return some years and fix our eyes upon the early settlements along the Monongahela. Even before the ill-directed and ill-fated expedition of British and provincials, under Gen. Braddock,[5] some few habitations of whites had begun to appear along our streams; and in one of those rude cabins appeared and smiled Ellen Aylworth. Like a rose in the desert did I see this beauteous flower bloom. At that early time what little of society was to be seen, presented with much of kindness, a stern inflexibility of purpose, and with that tenacity, a promptness of action, which crowded events upon events. As the thunders of the Revolutionary War began to be hoarsely heard beyond our mountains, Ellen rose to womanhood, but with her rose another— Saul Garvin. This young man was light of form and fleet as the deer on the hills. In person and in natural manners, never have I since beheld the equal to Saul Garvin. Ellen Aylworth was life and beauty personified— Garvin personated the times; he was serious but his heart was warm. In other regions and stations Ellen would have shone amongst the gay, the gifted, and the great. They were two whom nature placed together and forced to love, yet their hearts were not alike moulded—they were not destined to tread happily over the rough paths before them.

It always makes me smile when I hear the pride of wealth named, for never have I seen any rank—for I have seen all ranks—where this spirit was not equally active; and Saul and Ellen felt its malign influence. Saul was under the care of an aged uncle, and was pet and heir.— The wealth he was heir to, would not have been sufficient to fit out one of our modern dandies to pay a visit to the lady of his choice; but it was the most extensive fortune in the woods of Chartier, and no man ever prided himself more on his superior wealth, than did

5. On July 9, 1755, General Edward Braddock led an army of colonial militiamen and British regulars into a trap laid by the French and Indians along the Monongahela near present-day Pittsburgh. Darby refers several times to this fatal blunder in "Mark Lee's Narrative."

old Hall Kent. With the nephew, the lovely Ellen was worth all the money on the earth, but with his uncle her beauty and innocence was just worth nothing. The opposition of his uncle was only one of Saul's vexations. Another poor young man beside Saul, saw and sought Ellen. If Saul sighed, Tielman Wells laughed, and Ellen, in the gaiety of her young heart, laughed with Tielman, and the poison of jealousy rankled. Saul was deceived and so were the neighbors, and so was the joyful old Hall Kent.

So went on affairs for some days, and even weeks, and all the folly, passion, and extravagance of the world was acted on Chartier. Tielman Wells owned two horses, a saddle, bridle, watch, two rifles, and had by him ten old Spanish dollars, and had also the full approbation of Kent. Ellen smiled when his name was mentioned, and looked grave at the sound of that of Saul Garvin. Some wise one remarked, how wonderful it was that Ellen could choose such a skipping raccoon as Tielman Wells, and reject Saul Garvin. Such a preference would have been wonderful if made, but though then young, I thought I could see as far into a mountain as any one, and if no one else did, I saw the true state of the case.

On a snowy winter morning, about ten miles from where the fine city of Pittsburg now stands, with my rifle on my shoulder, I was traversing the Chartier hills in pursuit of game. Amid the loaded branches and falling flakes, I dimly saw the figure of another hunter, crossing the slope of the hill below me, and quickly perceived it was Saul Garvin. Though in the untrodden woods he was slowly bending his steps towards the house of the father of Ellen. He was arrested by "Saul, holoa! You are too late; Ellen is gone to Pitt, with Tielman."

"Gone to Pitt with Tielman Wells," replied the young man, as he approached where I was standing.

"Gone! yes," rejoined I, "and do you turn your course and flirt with Jane Sparkle, and Ellen will come to her senses."

He looked me steadily in the face, and with a visage no man but a hunter or an Indian could ever assume, pronounced as he cordially shook me by the hand, "Farewell! I have a longer walk to make than to Jonathan Sparkle's."

I was rather puzzled to know whether to laugh or be serious, but the latter mood prevailed, as I was in a moment alone in the forest. The look of Saul fixed on my mind and made an impression I could neither account for, nor for a moment forget. Thus silently impressed (for I communicated the circumstance to no person), two days passed, and on the third morning the dreadful report was spread, that Saul Garvin was missing—and murdered as was supposed.— With all my speed I hasted to the house of the distracted uncle, and revealing the meeting with Saul, led his uncle and a body of armed men to the spot on which we had met. A rapid thaw had laid most part of the hill sides and tops naked. For some miles we found tracks which we supposed to be those of the lost hunter, traced on the snow remaining in deep vallies and the northern slope of the hills, but reaching the Ohio, all further search was useless.

Twenty-three years the fate of Saul Garvin remained wrapped in mystery, and within the same twenty-three years the grave had closed upon the regret of the heartbroken uncle and the wasted form of Ellen Aylworth, and had also been marked by the ever to be lamented massacre of the Moravian Indians on Tuscarawas.— Seventeen years had the dust of the victims mingled with their parent earth, when the tardy justice of the United States recalled to their property and homes the remnant of the Christian Delawares. Seventeen years had I never dared to visit the desolate spot, where so· often I had met the warmest welcome; but when I learned that the poor surviving wanderers were to return, I determined to meet them. A young man of this neighborhood, of the name of Thomas Car, was appointed to meet the returning Indians with a supply of provisions, and with him I went, and was present when the aged chief and his forlorn band reached the scene of murder. Time had greatly changed us both, he knew me not, nor did I disturb his deep reflecting sorrow by any renewal of our acquaintance—to confess the truth, shame withheld me; as a white man I felt a share of the dreadful wrong, as did the plain uneducated young man beside me, though both were innocent of the deed. We silently followed the aged chief as he led a grandson by the hand, and pointed out where the houses formerly stood. Many of them had been supplied with cellars, at one of which much longer than the others, the old man stopped. It is seldom an Indian man weeps, but I saw the tears fall from his furrowed face into the hollow space. His feelings were for a few minutes that of human nature, but he seemed at some sudden thought to remember he was an Indian Chief, as he turned round and addressing his grandson observed firmly, and in English, "The grave of your father."

The young man sat down upon the slope of the cellar, drew his blanket over his head and remained silent, though his heaving breast showed strong agitation.

With more of sorrow than anger, the chief fixed his eyes on me, and in very good English observed, "I hope you were not here at—"

"Not at the time you mean," I solemnly interrupted, "thanks be given to the great spirit who rules over Indian and white."

"You are here now," replied the chief, "therefore I believe your words—seventeen years of reflection must keep him away who was here then."

"Many of them," I replied, "have gone to their judgment seat."

"Some have, I too well know," mournfully rejoined the chief, "and terrible was their departure."

"Were you present?" I rather hastily demanded.

"Not when and where you mean," emphatically replied the chief, "and thanks to the great spirit that I was not."

We all remained silent for some time, when the chief again addressing me, demanded "Were you ever acquainted with a white man of the name of Saul Garvin?"

"I was, and well, what of him—do you know—"

"Too much," replied the chief— "too much for the white—warriors I will not call them—who dyed this place in blood. Sit down on this bank and you shall."

We sat down, and after a long pause the chief resumed: "Twenty-three years have the leaves of these woods been renewed and have again fallen, since hunting on the high hills towards the rising sun, I saw the smoke of a camp. The war-hatchet was then buried, and I approached the fire and by it found a young white man; he met and received me kindly, presented me some venison, and we feasted together. It was evening, and he invited me to share his shelter, and we slept together. In the morning I invited the white man to go home with me, and told him we could reach there when the sun rose so high, and I pointed to the south."

[the Moravian chief]

"Of what nation art thou?" demanded he.

"A Delaware, in your language," I replied, "and a Christian."

He looked in my face long and thoughtfully, and then spoke, "You are a christian; I think you do not deceive me—your place of residence, where?"

"Schoenbrun," I replied.

"Do you receive white men into your tribe?"

"Very seldom; they are only bad men who leave the house to live in the wig-wam."

"Not so always," said the young white man, quickly, and I was sorry for using the words *bad men*, and taking him by the hand, told him to come with me and we could talk more as we went on our way. He then shouldered his rifle and we set out for Schoenbrun, where, before we arrived, I was told by the young white man that he had been deserted by a young woman he intended to make his wife.

"Was she the only young woman of your tribe?" I asked.

"There were more," he replied, "but none like my ———" he did not pronounce her name.—I thought him so far foolish, but told him we had some girls with us who could make moccasins and leggins. His smile was that of an Indian and not of a white man.

Our white friend remained some days with us, and we were all well pleased with him. Our old friend, whom you call missionary, was also well pleased, and we agreed to receive him.

"As an Indian," said the young man, "and a christian will I live,"— and he fulfilled his promise.

"What name shall we give you?" I asked.

"Peter," he replied, and white Peter was soon after married to one of our young women, and over this grave—for grave it is—stood their house.

[Kingsley Hale]

"The house of Saul Garvin!" I interrupted.

"The house of Saul Garvin," repeated the chief, "and amongst that earth," pointing to the bottom of the cellar, "lie the bones of Saul Garvin, and that sorrowing young man is his son—but hear me. Peter was soon an Indian in dress, manner, and language, for he learned our tongue and became a chief and a wise one at our council fires. To the few whites who came to our villages, he was distant and reserved; and in his new character remained unknown to all, and particularly to one or two who had as a white man been his intimate acquaintance: 'They are strangers,' said Peter to me, 'I know them no more; they are here for no good.'"

[the Moravian chief]

The snow had melted and was gone; the winds of winter had passed, and the song of the birds was heard amongst the new born leaves and flowers; we were preparing our corn fields, when a swift messenger entered Schoenbrun and told us "Wingenund is coming." Though still a warrior, and as you say a Pagan, the face of Wingenund spoke peace, and his words were words of truth, therefore we were glad, and walked joyfully to the council fire. The chief arrived; his face was sad and his greeting slow and mournful. The warriors who came with him, waved their hands in kindness, but from their lips only escaped a few words, which faintly reached our ears. The clouds of fear passed over our hearts as the chief rose; we had never before seen Wingenund as we saw him now. — His dark eye fell upon us as his right hand rose and was stretched to the east:

"Do you see the wolf stealing from those woods to

tear the timorous fawn to pieces? His tooth crushes its bones; he drinks its blood and howls over the mangled limbs. You are the fawn—the pale faced warrior, no, the white murderer is crouching to spring upon and destroy you, your wives and little ones. Where are your arms? buried—trampled down so deep, you have not time to dig them up. You sit down, and one ear is deaf, you cannot hear your red friends call to a place of safety—no! but you can hear the white man say, sit still, you are safe. Hear me, children, rise quickly and fly with me. The white man smiles, but see, the sharp knife is under his blanket. Come with me."

Wingenund sat down, and we all sat still a long time. At last Peter rose. We were all astonished, for Peter was no great talker, but we were pleased, and our ears were open. The young chief spoke thus:

"Fathers, I am young and ought to hear what older and wiser men have to say, but your mouths are closed and I must speak. The word of our father comes from the great spirit, it is good. I was once a white man, and know that many white men are good men and brave men, who would not seek the unarmed to shed their blood; but I know there are bad white men, whose ears and hearts are deaf. These bad men say you are the friends of your red brethren, who have taken up the hatchet with their red-coat white enemies. The voices of the good white men are soft and cannot be heard afar, but the voice of the bad man is loud as the panther's scream.— Hear me, my friends, arise and go with Wingenund." Peter sat down and we were again all silent until Wingenund again rose and spoke.

"Tomorrow morning I return to my people."

One of our old men then rose and said, "Such as choose to go with our brother be ready."

We slowly and sorrowfully sought our houses and families. The wife of Peter was my daughter, her christian name was Anna, and that young man was then lying on his mother's bosom. Their house was over this grave. We entered and sat down. Anna feared, she knew why—wept, and fixed her face on that of her son.

"Anna," said Peter, "you must go with Wingenund."

He then told her what I have told you. "And you, Peter, will stay here if any others stay?" Anna was a woman, a mother, a wife, and a christian, but she was an Indian, and would not refuse to do what her husband and father desired her to do, and in tears and silence prepared for her journey.

In every house was heard the voice of distress. Some whole families concluded to fly from the danger, but too many said, "What have we done that we should fear the white men beyond the Ohio? To such as have come here we have been kind, and why should we fear? We worship the same great spirit with the whites, and are we not their brothers? We will stay and trust in God."

Morning came, and Wingenund departed with such as were to depart with him; but, as he was departing, he turned round and said to those who remained behind, "You say you are christians like the whites, why should we fear the white men? I tell you why you should fear many of these pale faces; it is because they are not christians;" and he was soon beyond our sight.— I do not tell you of the parting between those who went with Wingenund and those who stayed on Tuscarawas— I have no words."

[Kingsley Hale]

The old man remained silent for some time, when again conquering his feelings he resumed:

[Moravian chief]

I went with my child and grand-child, and silent and painful was our journey. It was not long that we were left to think upon our own miseries. The friends we left on the Tuscarawas were also not long left to tremble between hope and fear. The storm foretold by Wingenund, burst upon their heads. A body of men under a man of the name of Williamson, rushed upon the defenceless, the unarmed, and, as you have been told, unsuspecting people. This band came not as open warriors but still with words of peace upon their lips.

"You are unsafe here," said they to our brothers, and their wives and children, "you must come with us, and stay with us until the war-hatchet is buried."

"We are not afraid," replied the red men.

"The Shawnees and Mingoes will come and destroy you," repeated the whites, "you must come with us."

It was useless to resist, and the Indians were preparing to follow their enemies. Night passed, morning came, and they were told to prepare for death. The looks of too many of their savage enemies had forewarned them, and the sentence was expected—they were prepared. In prayers, hymns and tears, their last night was spent, and when the sound of death reached their ears, they bowed their heads, not to their murderers, but to Him whose decrees no man can reach.— They perished, and with them fell Peter, or Saul Garvin. A few young at the last moment, bounded from their captors, and though some fell in the pursuit, some escaped and brought us the dreadful tale.

One of the men who read death on the white faces in the night, spoke in our language to Peter, and asked him why he did not make himself known. "Because," said Peter, "I know some of those pale faces too well. To make myself known to them would do more harm than good. If they carry us into the settlements I will make myself known, and will do you all the good in my power. You received me when a bad man forced me from my own people. You have treated me kindly. I have lived and will die with you." He then desired to be left to himself, and was heard to speak no more to men; but his inward voice was turned to the great spirit, and in the morning his blood sunk into this ground, and from it cried to the great spirit for vengeance, and was heard; the messenger of revenge you will soon hear named. Nearly one hundred unarmed human beings were here murdered and their bodies left to moulder amid the ruins of their houses. The panther, when he tastes, is never satisfied with blood. He drinks blood and his thirst burns more and more fiercely. So did your people. The cries of the murdered had scarcely ceased to be heard along this river—the smoke of their houses had scarcely mingled with the clouds, when another party of white men came into our country, under Williamson and Col. Crawford.

Meeting no resistance in the first instance, these men

became bold; they advanced far into our country—approached the Sandusky towns—were met by warriors, defeated, and Col. Crawford and many of his men taken and brought bound to our council fires. Those who escaped death in the battle, or captivity, were scattered over the woods, exposed to the rage of the Indian to hunger and to wild beasts.[6] The bodies of many were made known to the Indian pursuers by the vultures' flight. Some did return to their homes, not to dwell upon the deeds of the brave. But let us leave them who fled and return to those who in bonds had to bear the punishment of the wrongs committed by others, and to writhe in despair at their own madness and folly.

I was, with my daughter and grand-child in Detroit, when we heard of the murders at this place. Anna clasped her babe to her bosom, and raised her heart to the great spirit—wept—was silent, and was daily wasting away, when the second news reached us, that Col. Crawford and some of his men were in the hands of the warrior Delawares. Anna was a christian, but she was an Indian; and I am an Indian, and will tell the truth. Over the memory of her husband, she was sinking towards the grave—the world of spirits, where she hoped to meet her Peter; but from the moment that the captive Crawford was named, I saw new life in her eye,—health seemed to return to her body, yet, not even her father suspected her purpose. Next morning Anna and her son were gone. I followed, but found her not until her terrible resolve was fulfilled.

Crawford, before the war-hatchet was dug up and dyed in blood, had been the friend of the Indians, christian and warrior; and when he was brought bound to the Sandusky, the Delaware warriors shook their heads and said, "We are sorry. This man was once our friend;

6. News of Crawford's defeat stunned the pioneers of western Pennsylvania. His men were indeed "scattered over the woods"—as Darby suggests through the story of Powers Osborne in "The Wedding." Crawford himself was taken captive, scalped, and tortured with fire before he finally expired. The dramatic act of Saul Garvin's widow, depicted by Darby, is pure fiction; in fact, Crawford's miseries were witnessed by Simon Girty, the white renegade, who "laughed heartily, and, by all his gestures, seemed delighted at the horrid scene." See Consul W. Butterfield, *An Historical Account of the Expedition Against Sandusky* (Cincinnati, 1873), 389.

he has not come into our country as a wolf, but as a man. How glad would we have been had he been killed in battle, or escaped, and we had Williamson in his place." Thus spoke Wingenund and many more, but other voices were heard breathing revenge. The prisoner was brought before the Delaware warriors, and many looked to the ground, and all were long silent. At last an aged warrior rose, and his words fell like the edge of a heavy and sharp hatchet.

[the Delaware Warrior]

Where are we? Over the mountains, from the land of our fathers; and why are we here? The whites have driven us from stream to stream. We have often smoked with them the pipe of peace, the white man putting his foot on and pushing deep into the ground the war hatchet, and holding in his hand the speaking book, saying, "Red men you are blind, but here is the word of the great spirit, which will make you see." We answered, we have not learned to hear with our eyes. "Well, we will send you some of our black coats," replied the white men, "and they will speak to you from the book." Well, the black coats came to us and told us, that the great spirit loved peace; that we must not only bury, but burn the war-club and hatchet.

Well, some of our people loved peace, and believed the words which they were told came from the great spirit. These people broke and buried the war-hatchet and put the handle into a hoe and made corn. Their children laughed and played, and their wives sang songs from the speaking book.

But when the white men dug up the hatchet between themselves, our christian brethren were afraid; but the black coats told them not to be afraid. Some of our old men went to them, and told the Christian Indians, "We know there are some good white men, but they are few. We know there are bad white men, and they are many; and they rule the good men. There is no faith to be put in their words. They will shake the Indian by the hand and say, 'Brother, friend, my own brother, my own friend,' the next moment his knife is in your heart, and your wigwam is in flames; your wives and little ones lie beside you bleeding. Trust them not—remember I have told you—trust them not."

The voice of the Indian was not heard by many, and where are they? You are silent.— Praying men, singing women, and laughing children, murdered by such white men as the man we have here. "He must die," cried many voices; but the chief continued:

Hear me—why must he die? It was not him who murdered the praying red men and their families. If we had Williamson and any of the other cowards who committed the murder, they ought to die. We ought to become wiser—

[the Moravian chief] The chief was here stopped, and the whole council surprised by the sudden entrance of a young woman with a young child. She stood sometime with the child, and then laying him down in the centre of the room, cast a look of fury on Col. Crawford, and cried, in a voice which made even the oldest warrior tremble, "My child—his father a white man. His bones lie with the bones of our people. My child, he has no father." She then rushed from the council house, leaving the child on the ground at the feet of our warriors.

A silence, only broken by the groans of the prisoner, continued some time, but it was the silence of death; the child at first amused with some object, missed at length his mother, and looking round and not finding her, screamed aloud, and his cries were the cries of death. The voice of mercy was as the voice of the fawn to that of the panther. Col. Crawford perished in the flames, and—

[Kingsley Hale] "His spirit," after a long pause, continued the chief "has been many years gone to meet the God of red and white men. Peace now reigns over red and white men. We have returned to rebuild our cabins, and again plant our fields.— Our tears are dried, and we can sleep without fear and rise to joy and plenty; our children can sport in safety and our women sing the song of gladness along the Tuscarawas."

# A Chronological List of Darby's Narratives and Sketches

"The Vendue" (*Saturday Evening Post*, June 6, 13, July 18, 25, 1829; *Casket*, July, August 1829).

"The Sioux Chief" (*Saturday Evening Post*, October 10, 1829; *Casket*, October 1829).

"The Indian Trader; or, James Bolton, of Orange" (*Saturday Evening Post*, November 21, 1829; *Casket*, December 1829).

"Ann Dillon" (*Saturday Evening Post*, January 9, 1830; *Casket*, February 1830).

"Caroline Marlow" (*Saturday Evening Post*, March 13, 1830; *Casket*, March 1830).

"Clement Meyerfield and Clara Ismeana" (*Saturday Evening Post*, May 22, 29, June 5, 12, 1830; *Casket*, May, June 1830).

"Pulvinara, or Tales of the Pillow" (*Saturday Evening Post*, September 4, 1830; *Casket*, October 1830).

"Landerman" (*Saturday Evening Post*, September 25, 1830; *Casket*, October 1830).

"The Will, or High Expectations Modified" (*Saturday Evening Post*, October 9, 1830; *Casket*, November 1830).

"Henry and William Nelson" (*Saturday Evening Post*, November 13, 1830; *Casket*, January 1831).

"The Inquest of the Dead" (*Saturday Evening Post*, December 4, 1830; *Casket*, December 1830).

"The Spirit of the Potomac" (*Saturday Evening Post*, December 18, 1830; *Casket*, February 1831).

"The Shipwreck, or The Hermit of the Rock" (*Casket*, April 1831).

"The Hunter's Tale; or, Conrad Mayer and Susan Gray" (*Saturday Evening Post*, December 17, 1831; *Casket*, December 1831).

"The Moravian Indians" (*Saturday Evening Post*, May 25, 1833; *Casket*, May 1833).

"Ellery Truman and Emily Raymond, or The Soldier's Tale" (*Saturday Evening Post*, December 7, 1833; *Casket*, December 1833).

"The Three Wishes, a Dream" (*Saturday Evening Post*, February 8, 1834; *Casket*, February 1834).

"Mark Lee's Narrative" (*Saturday Evening Post*, May 24, 1834; *Casket*, July 1834).

"Cyrus Lindslay and Ella Moore" (*Saturday Evening Post*, June 28, 1834; *Casket*, August 1834).

"The Great West" (*Saturday Evening Post*, July 12, 1834; *Casket*, September 1834).

"Julia Gray, or The Orphan" (*Saturday Evening Post*, August 2, 1834; *Casket*, September 1834).

"Reminiscences of the West" (*Saturday Evening Post*, December 20, 1834; *Casket*, December 1834).

"Letburn Parkman, or The Maniac" (*Saturday Evening Post*, January 10, 1835; *Casket*, February 1835).

"Stillwood" (*Casket*, January 1835; *Saturday Evening Post*, February 14, 1835).

"Gilbert and His Family" (*Casket*, April 1835).

"Alloys Ganoway" (*Saturday Evening Post*, August 1, 1835; *Casket*, August 1835).

"The Consumptive" (*Saturday Evening Post*, August 29, 1835; *Casket*, October 1835).

"Ann Eliza Glenn, a Tale of Wyoming" (*Saturday Evening Post*, September 12, 1835; *Casket*, October 1835).

"Lydia Ashbaugh, the Witch" (*Saturday Evening Post*, January 16, 1836; *Casket*, January 1836).

"The Wedding" (*Casket*, July 1836).

"The Unknown" (*National Atlas*, July 31, 1836; *Casket*, September 1836).

"Ashgrove, or The Return" (Washington, Pa., *Reporter*, n.d.; *Saturday Evening Post*, October 27, 1838; *Casket*, November 1838).

# III. Geographical Writings

# *Textual Note*

The relative neglect of Darby in our own time has made it desirable to represent the nature and scope of his geographical writing, the best of which appears in the three works cited here. Recent reprints (by Scholarly Press) have now made these titles readily available; hence, the passages offered here have been selected for their readability and general interest rather than their geographical significance. These extracts nevertheless reveal fundamental aspects of Darby's thinking about nature, western settlement, the benefits of civilization, and the lessons of history. Here and there, the reader may recognize lines quoted earlier in the biography; some repetition was unavoidable, since these excerpts contain a few of Darby's most memorable observations.

His geographical writings generally received more careful editing by his publishers than did the magazine tales. The reader is therefore troubled less frequently by cumbersome constructions or ambiguous usage. Emendations for this reprinting are infrequent and insignificant. Place names, however unorthodox the spelling, have been preserved intact. As noted earlier, Darby had no pretensions as a stylist, and given the fact that his formal education ended in early childhood, we should see his prose as a remarkable instance of his passion for learning and self-improvement.

# A Geographical Description
## of the State of Louisiana

The text for these excerpts is the "enlarged and improved" second edition of 1817, published in New York by James Olmstead. Darby enlarged on the first edition principally by adding chapters on the state of Mississippi and the Alabama Territory, the coastal areas of which the author had surveyed. The *Description of Louisiana* initially accompanied Darby's giant map of the state; a smaller version of the map was inserted into the second edition. Without question, the book and map together constitute the author's most significant single contribution to American geography. His was easily the most complete and accurate topographical representation available in 1816; his precise prose account of the state's physical features, designed to vanquish the misconceptions of eastern readers, still possesses geographical and historical interest. In his "Explanation of the Map," appended to the second edition, Darby forthrightly identified the principal achievement of his study: "I am the only man who ever did attempt and execute a survey of the Sabine and its confluent waters."

from *Chapter IV, Statistics of the State of Louisiana:
Attacapas and Opelousas*

*Prairie Calcasiu.*—This extent of grass is from N. E. to S. W. fifty miles long, and twenty miles wide, having more than 640,000 acres of land.

The soil along the east border, on the Nezpiqué, is of second rate quality, its surface is rather more waving than prairie Mamou. Along the Mermentau the prairie exhibits gentle swells, which relieve the eye from the dull monotony of the unvaried plain. The west margin of the upper lake in the Mermentau is a most beautiful slope, rising with gentle acclivity twenty or thirty feet, and falling by a more imperceptible declination into the general expanse of the prairie. Some handsome situations for building are found here. The lake upwards of a mile wide and more than six long, spreading under the eye, diversified with one or two small islands covered with trees, the interminable expanse bounding the view on all sides, except limited and relieved by the woods on the Mermentau to the north, or the small clumps of wood scattered in pleasing confusion in every other direction.

Below this lake, timber ceases on both sides of the river, which here swells to 400 yards wide, bordered by a very narrow bank of shells on one side, and the impassable morass on the other, having depth of water for large vessels. All possibility of settlement ceases. Twenty miles below the Little lake, following the stream, the river opens into another ten miles wide by twenty long. The channel less deep, and more uncertain. At the west extremity of this lake, the Bayou Lacasine comes in from the N. W. The latter Bayou may be considered the drain of this prairie. Like the other prairies of Opelousas and Attacapas that are bounded by the sea, its marine extremity is an impenetrable morass, except through the rivers. The Lacasine has no wood on its banks, many miles above its junction with the lake; its channel is deep enough for large vessels. After wood commences, the adjacent prairie rises above the marsh. Some good soil is found, but not of large extent. The forest timber are oaks of several species, pine, ash, hickory, cypress, and tupeloo. The dwarf trees on the higher lands are dogwood, and whortleberry. The east branch of this Bayou remains navigable after the wood ceases. The other branches dwindle to gullies, on leaving the prairie. Not more than twenty or thirty families could be comfortably fixed on this Bayou. Most of the land remains to the United States, though three or four claims are surveyed on the east side. Between the Lacasine and the pine woods on the north, and the Mermentau river on the east, the face of the earth exhibits an expanse of grass, interrupted only by an occasional clump of oak or pine trees, that resemble isolated savages, trembling alone from age to age. After passing Lacasine, the same monotony again re-assumes dominion. The winds breathe over the pathless waste of savannah. The wild fowl is seen flitting, or the deer skimming over the plain. The clouds of heaven close the picture on the south; while fading in the horizon, the far seen woods, raise their blue tops between the prairie and the sky, in every other direction.

At any considerable distance from the woods, the land is sterile, and even near or in the forest, is of a very unpromising texture. I am led to think this region healthy; not many of the causes that produce destructive miasma exist here. The truth of this position is proven by the few persons that have settled on either part of this prairie. Grazing will, it is most likely, be the prevailing pursuit of the inhabitants of this part of Louisiana. In almost every place west of the Teche and Vermilion woods, many inducements must operate to give that current to their employments. But as the population of the banks of the Mississippi increases in numbers, and creates an accumulated demand for beef, butter, tallow, hides, and cheese; shipments of those articles will be made directly from the mouths of the Atchafalaya, Vermilion, Mermentau, Calcasiu, and Sabine rivers. Salt can be manufactured to any amount, in many places near the coast; which will render that necessary article cheap. It may not be irrelevant to remark here, that nature and art will combine, to render the banks of the Mississippi the peculiar seat of abundance. The inexhaustible stores of provisions from the northern waters, combined with the boundless pastures to the west, give a facility and certainty of supply no where else found, in so eminent a degree, on our globe. (pp. 110–12)

\*      \*      \*      \*

From the mouth of Courtableau to the head of the Cow island, the breadth of the overflow between the Atchafalaya, Opelousas, and Attacapas, is about eight miles wide. This space is an immense lake, for many months; the currents of the smaller Bayous are lost in the maze, and only remain distinguishable by the openings of their channels. The many lakes that mingle with the outlets of the river, and with each other, render this region most inconceivably intricate. It is with the utmost difficulty that the real channel of even the river can be distinguished from the number of outlets and inlets, that wind in every direction. The forest trees are indicative of an inundated country; such as swamp white oak, indented leafed red oak, bastard paccan, white wood, persimmon, cypress, though not abundant, some species of the thorn, a species of the honey locust, and other aquatic trees. Below the head of Cow island, on spots along the margin of the river, (mostly on the right bank,) which are above overflow, the quercus, sempervirens, or evergreen oak, begins to appear; some of the candleberry myrtle fringe the shores with their deep green foliage and impurpled fruit; here also appear spots of cane, but of no great extent, the narrow selvage of high land quickly receding into the dead overflow.

To have an idea of the dead silence, the awful lonesomeness, and dreary aspect of this region, it is necessary to visit the spot. Animated nature is banished; scarce a bird flits along to enliven the scenery. Natural beauty is not wanting; the varied windings, and intricate bendings of the lakes, relieve the sameness, whilst the rich green of the luxuriant growth of forest trees, the long line of woods melting into the distant sky, the multifarious tints of the willow, cotton, and other fluviatic trees, rendered venerable by the long train of waving moss, amuse the fancy. The imagination fleets back towards the birth of nature, when a new creation started from the deep, with all the freshness of mundane youth. (pp. 135–36)

*　　*　　*　　*

A fact which may not be irrelevantly related here, will suffice to show, how slowly changes in the face of nature are effected by water.

Above the efflux of the Fusilier, and nearly opposite the Derbane, the marks of overflow are at the distance of a mile from the bayous, four feet high on the trees; yet there are in this dreary waste, six or seven of those little mounts, or barrows, found over almost all America, and the north of Europe and Asia.

In the year 1808, when first discovered by the author, their summits were still sufficiently elevated to be beyond the reach of overflow, and covered with timber, indicative of high land; such as black oak, sassafras, and ash, but more particularly three different species of vegetables were found, never known to occupy lands subject to annual immersion, viz. black gum, holly, and the muscadine grape vine. The latter, like the large cane, though found near the margin of the inundated lands, is never found within their limits.

Those mounts are about seven or eight feet higher than the water mark on the trees; and are scattered to some distance from each other, without regular arrangement.

The alluvial lands on the Teche are six miles distant, most of the intervening space liable to inundation. The spot where the barrows are found is a cypress swamp, a drain of which passes the space occupied by those sylvan towers. One of the smallest of them is broken by the action of the water. When we reflect upon the length of time necessary for such piles of earth, after their erection, to assume the antique form they exhibit at present, and to cover themselves with timber suitable to their exemption from overflow; the seeds of which timber must have been translated casually from a considerable distance; we are struck with the conviction, that many ages past the adjacent country was nearly in the same state that it is at present. Many theories concerning the causes or intent of such monuments, are much weakened by the situation of those now treated of.

Not even a village of savages could have existed throughout the year, within several miles of this place. The spot where they are situated, is more dreary and sunken, than most other parts of the adjacent swamp. There is much reason to doubt the correctness of the opinion, that those elevations were erected for either temples or dwellings; the probability is much greater, that they were cemeteries raised on the field of battle, containing the bones of the ancient warriors of Louisiana.

Human pride has every where erected monuments to perpetuate the crimes, follies, and miseries of mankind; monuments, themselves, perishable as the hands that built them. Whether the marble of Greece, the porphyry of Egypt, or the clay of the Atchafalaya; time sinks them all to eternal ruin. The pyramids of the Nile; and the barrows of the Mississippi, attest alike, the weakness and evanescence of human greatness. (pp. 156–57)

### from *Chapter VII, Statistics of the State of Louisiana:* *General Observations on the Climate of the Delta of the Mississippi and Adjacent Countries*

In an inquiry into the influence of the climate of Louisiana upon the health of the inhabitants, to complete the investigation, it will be necessary to establish its effects also upon the mental faculties of persons born within the sphere of its influence. This section we enter upon with a feeling of pleasure. The people of the United States will receive with equal satisfaction, a detail, that when admitted as correct, must lessen the prejudices that accident and design have engendered to widen the moral distance between them and their fellow citizens in Louisiana. To an ingenuous mind, nothing administers more solid gratification, than to find man more amiable than expected. The noble enjoyment arising from the exchange of sentiment between enlightened minds, is one of the greatest privileges that reason has accorded to man. To open new sources of this sublime fruition, is conferring a benefit on human nature.

The character of the Creole of Louisiana may be drawn in few words. Endowed with quick perception, his faculties develope themselves at an early age; if found ignorant, it is not the ignorance of stupidity, but arising from an education under circumstances unfavourable to improvement. Open, liberal, and humane, where he is found inhospitable, it is the fruit of a deception he dreads, and to which his unsuspecting nature has led him to be too often the victim. Mild in his deportment to others, he shrinks from contention; a stranger to harshness, his conduct in the pursuits of life is marked by kindness. Legal disputes, that seem to form part of the amusements of the people of some other parts of the world, are instinctively avoided by the Creole. His docility and honesty secure him from injuring others, and he enters the temple of justice with reluctance to demand reparation for his own wrongs. Sober and temperate in his pleasures, he is seldom the victim of acute or chronic disease. His complexion, pale but not cadaverous, bespeaks health, if not a vigorous frame. His strongly speaking eye, beams the native lustre of a mind, that only demands opportunity and object to develope all that is noble and useful to mankind. If the Creole of Louisiana feels but little of a military spirit, this apathy proceeds not from timidity; his ardent mind, light athletic frame of body, active, indefatigable, and docile, would render him well qualified to perform military duty, should this part of his character ever be called into action.* The peal of national glory was never rung in his youthful ear. One generation has arisen since Spain held this country, and noble was the germ that retained its fructifying power, under the blighting influence of that government. Louisiana has escaped the galling and torpid yoke; its inhabitants will share the genius and freedom of the empire in which they are incorporated.

The cordiality with which the Louisianians hailed their introduction into the U. States government, has received a check from the misconduct of too many Americans. The moment the change was effected, an host of needy adventurers, allured by the softness of the climate, the hopes of gain, and inflated by extravagant expectations, spread themselves along the Mississippi. Many men of candid minds, classical education, and useful professional endowments, have removed and settled in Louisiana; but some without education or moral principle, prejudiced against the people as a nation whom they came to abuse and reside amongst. Too ignorant to acquire the language of the country, or to appreciate the qualities of the

---

*This part of the work was composed at Opelousas, and read to several persons, in the month of October, 1811. How far the author estimated correctly the character of the Creoles, and the consequences of invading Louisiana, intermediate events have amply explained.

people, this class of men have engendered most of the hatred existing between the two nations that inhabit Louisiana. The evil of national animosity will gradually subside, as a more numerous and orderly race of people become the improvers of the public lands.

The dark side of the Creole character may be considered impatience of temper, and a propensity to licentiousness when in the possession of wealth.

\*     \*     \*     \*

I have reserved, to close the subject, the examination of that part of the human species, whose moral character has, in every civilized region of the earth, and in all ages, most deeply influenced that of man. It needs no other criterion to judge of the rank that nations may be entitled to occupy in the scale of civilization, than the state of their women.

The women of Louisiana are, with few exceptions, well formed, with a dark piercing eye. Their movements bespeak warmth of imagination, and a high flow of animal spirits, whilst their features indicate good nature and intelligence. Tender, affectionate, and chaste, but few instances of connubial infidelity arise from the softer sex. With too often example to excuse, and neglect to stimulate, the most sacred of human contracts is fulfilled on their parts with a fidelity that does honour to their sex. In all parts of the earth, and in all ranks of society, women are more virtuous than men. From some cause that operates every where, the moral sense is more deeply felt, and more uniformly obeyed by women than men: more temperate in their enjoyments, their passions are more under the guidance of reason; decent in their deportment, they continually counteract the predisposition in man to vulgar sensuality.

As wives, sisters, or mothers, the Creole women hold a rank far above their apparent means of education. Frugal in the expenses of life, they seldom lead their families into distress, by gratifying their pleasures or pride. Rigid economy, that may be called a trait in the Creole character, is more prominent in the conduct of women than in that of men. Very seldom the victims of inordinate desires in any respect, their dress is regulated by neatness, decency, and frugality.

That this picture is neither the effect of a warm imagination that delights in clothing objects in false colours, or that of flattery, will be admitted by generous, candid and observing men of all nations, who have had the honour to possess the only means of forming a judgment—converse and acquaintance with the objects of the inquiry. If the women of Louisiana are found deficient in mental endowment, the reason is obvious:

want of the means of acquirement. But the minds of the Creole women, remarkably active and tenacious, are much less ignorant than is generally supposed. Should a general taste for reading be infused into society, if a judgment can be formed by the strength of mind, intuitive perception, and clear discrimination evinced by the fair of Louisiana, their rank in the scale of intelligence will be respectable, if not exalted.

At this moment, politeness, ease, hospitality to strangers, tenderness to their relatives, and indulgence to their slaves, attended by a mild unobtrusive decency of deportment, mark the conduct of the Creole women. Exceptions may be found, but the general outline is just. (pp. 270–78)

# The Emigrant's Guide
# to the Western and Southwestern
# States and Territories

The text is the 1818 edition, published in New York by Kirk and Mercein. Darby produced two maps for the volume: a map of the United States (roundly criticized by reviewers) and a map of the Gulf Coast near Mobile. As comparison reveals, the contents of *The Emigrant's Guide* overlap the *Description of Louisiana*; over half of the book, in fact, treats Louisiana, Mississippi, and the Alabama Territory. Darby also padded the work with government documents, proclamations, remarks on European geography, and a chapter mainly about olive trees. His descriptions of Ohio, Kentucky, and Tennessee, as well as those of the Great Lakes regions, are largely statistical, based more on library research than direct observation. However, this is not the case with his account of Pittsburgh, which conveys a graphic picture of that burgeoning industrial center in 1815. Probably the most interesting section is his "Advice to Emigrants"—a homily on the risks and benefits of westward movement, based in part on the author's own painful experiences.

## from *Advice to Emigrants*

With all the maps and descriptive works that can be procured, no emigrant ought ever to purchase land, or make arrangements for permanent settlement, before viewing the place where his purchases or settlements are to be made. The most that reading can do in favour of the emigrant, is to prepare his mind with more clear ideas of the means to form a judicious selection. Another necessary precaution is, to always distrust the information of persons offering lands for sale. Inquiries ought to be carefully made respecting the seasons, climate, diseases; and made as much as possible from persons whose interests are not engaged on the side of a too favourable representation.

Most men on arriving in the United States, expect too much. Perhaps the only essential advantages offered, are the security of person and property, and the cheapness of land. It demands excessive labour, severe economy, and exemptions from extraordinary accident, to succeed in a newly settled country; and it demands the permanency of this suit of labour, prudence, and favourable circumstances. In West Pennsylvania, West Virginia, in Kentucky, and in Ohio, where the establishments have continued a sufficient length of time, the emigrant will find inumerable instances to stimulate his exertions. Many persons of good character and intelligence, reside there at this moment, who have crossed the Aleghany mountains within the last thirty-five years, "*the world before them and Providence their guide,*" who now repose in ease with flourishing families around them. The emigrant who now traverses those mountains has no savage warfare to appall him. The first race of men who entered those wilds smoothed the path for their successors, often at the expense of their lives. What once demanded almost superhuman bravery, now only demands persevering industry, and honest sober habits. A great proportion of the entire number that now reside in the Ohio and Mississippi valleys, are persons who carried with them little more than experience in their respective pursuits, and who have created their fortunes by their labour and ingenuity. This is not particularly the case with agricultural men; it forms the basis of the private history of all classes of society. The consequence of necessary exertion has been to form a race of active, laborious and enterprising men, equal to any that the world has produced. The vast scale upon which the merchants and farmers of the Ohio and Mississippi valleys perform their operations is indeed expansive. It will be seen that from Pittsburg to New-Orleans is about two thousand miles, and also half that distance from the junction of Ohio and Mississippi to the latter city. Yet great numbers of the farmers are their own factors at so distant a mart.

The commencement of their course of business is, properly speaking, in autumn, when their grain is put in the earth. As soon as seeding is finished, preparations are then made for converting into flour or whiskey their small grain, in fattening their pork, and, in fine, collecting for market the various staples, and in building boats for the transportation of their property down the rivers to the mart of sale. In this manner autumn and the beginning of winter is consumed. As soon as the spring freshets open the rivers, these navigators commit themselves and the fruits of their fields to the current, and in due time float to Natchez or New Orleans; dispose of their cargoes, and purchase a horse, and return home by land. Every one is anxious to complete his voyage in time to return to his farm by harvest, which two-thirds effect.

The same routine is again pursued, and thus while some members of a family are as high as the 41° north lat. tilling the ground, others are distant eleven degrees of latitude disposing of their joint property. So easily do men accommodate themselves to the operations of this wide field of action, that many who, in their native country, considered thirty or forty miles a very serious journey, converse familiarly upon a voyage of two thousand miles from home, and a journey of twelve hundred to return.

One of the most valuable results of the distant voyages and journeys made by so many, is the infusion into society of an extent of topographical knowledge no where else known on earth. There is no exaggeration in declaring that no people in the civilized world can, in an equal population, produce so many men who possess general and detailed knowledge of a space so immense.

Most of the traders are well disposed to communicate to strangers such information as they possess, and very few are disposed to deceive. They are, in fact, a bold, open, intelligent, and candid body of men. They are the links of a chain of extensive communication. Like all other men of the west, the farmers and traders have a peculiar apparent carelessness of manner, which strangers, even from the eastern side of the Aleghany, are very apt to mistake for want of attention to those who address them. The fact is far otherwise: often when the traveller is thus thrown from his guard, he is in the presence of a man who penetrates the inmost recesses of his soul, and who will recount to his companions the very train of reflection passing in the mind of the stranger during this inspection.

One of the greatest and most fatal faults committed by Europeans when in this, as they term it, verge of civilized life, is undervaluing the inhabitants. It is in many respects a very natural result of the accounts published and read in Europe. One traveller, who, between New-York and Philadelphia composed two large volumes on the general characteristics of the United States, very gravely informs his readers, that in receding from those cities, the scale of civilization lowers, until upon the Ohio and Mississippi the savage state commences. Though it can hardly be supposed that many persons can be dupes to such representations, yet, from their tenor, prejudices must follow in the minds of those who read them. It is against the consequences of such ill-judged colouring we now wish to guard the emigrant. These calumnies do very little harm to the objects; but are extremely mischievous to those who travel the interior of the United States under their influence. Hatred and contempt are plants of easy growth, and very difficult to eradicate when once rooted in the human heart.

With a good personal character and suavity of manners, it is almost impossible for any man to reside three months on the western side of the Aleghany mountains without finding employment sufficient to provide for his subsistence. Every man who carried with him those requisites will find a kind welcome every where, and a disinterested advice in most intelligent men he meets.

All trades are wanted, especially those necessary for the supply of the most pressing wants of new settlers, such as carpenters, masons, smiths, wheelwrights, tanners, curriers, tailors, shoe-makers, batters, saddlers, and cabinet makers.

Mere labourers, however, who possess no handicraft, are as certain of employment as any class of men; so great is the task of clearing land, ploughing, sowing, reaping, threshing grain, and other business of husbandry, that all men can find work, who are disposed to gain an honest and virtuous subsistence. To the latter, and to common journeymen mechanics, we desire to point out a rock, that, as they value future reputation and happiness, must be avoided;—it is the idle waste of Saturday afternoons in play, or what is worse, in the grog shop. Why this part of time should be so unprofitably thrown away as it is, it would be difficult to explain; but the facts are too numerous to be doubted. Thousands who labour, attentively, through five and a half days, lose the fruits of their toil and their peace of mind in the other half, and rise upon the morning of the true day of rest much more inclined to repeat debauch, than to perform the sacred duties, that all laws, divine and human, have imposed; duties, that to perform is to secure the highest enjoyment of which our nature is susceptible.

Let the poorest young man of from twenty to thirty years of age, who finds himself in the theatre we have under our view, only turn his eye towards the different members of society, and at every glance he will find men in different circumstances, who, at a similar age with his own, had no other patrimony but wealth of body and mind, and who experienced no other good fortune but the effects of well-conducted labour. If from Europe, he will find nothing of the hauteur of high life, towards men who are engaged in honest industry. He is there relieved from that depression of heart that arises from contumely, "the proud man's scorn." Treated as a party to a fair contract, and not as a dependant, his mind expands, his nature becomes daily more exalted, and feelings and virtues arise in his soul of which he had no previous conception.

Many will say that these observations can only apply to the people of the states and territories where slavery is prohibited. That is, however, not the fact; a residence of sixteen years in places where slavery is prevalent, enables us to contradict a general expression, that in such places, whites, performing manual labour, are confounded in the moral estimates of the people with slaves. Though less respect is certainly paid to useful labour in the slave states than where all the duties of life are performed by the whites; yet the distance between the two races of men are in all cases immense. So deep, profound, and inveterate is the feeling on that subject, that not any where in the United States, can property, sobriety, intelligence, and every other advantage, except colour, raise in public opinion a man the most remotely allied to the African, to a rank equal to the meanest white. Any person who resides a few years in Louisiana will be witness to some very remarkable exemplifications of this innate contempt for all those whose affinity involve them in the contumely heaped upon men degraded by slavery.

Some of the most wealthy planters in the two states of Louisiana and Mississippi have made their outset as mechanics. They are now respected, in exact proportion as their conduct merits. There exists no country where skilful mechanics, particularly carpenters, blacksmiths, millwrights, bricklayers, and tanners, have a more fruitful field before them than in Alabama, Mississippi, and Louisiana. If attentive to the duties of their professions, they incur no risk of being confounded with any class of men but the virtuous and the honest.

One circumstance alone can degrade the white man in any part of the United States, to a level with the slave; that is his own moral dereliction. It is this source from which has flowed almost all the supposed contempt experienced in the southern states by labouring men.

The whole of these admonitory lessons may be summed up in few words; that with caution, temperance, honesty and industry, most men will not only secure competence, but wealth, in any part of the valleys of Ohio and Mississippi.

The lessons that can be given respecting health would be in great part a repetition of what has, or might be, said on the subject of wealth. There is one circumstance in the former case but little connected with the latter; that is, exposure to night air. In all places in the United States south of Tennessee, and in summer, in many places north of that state, night air is extremely deleterious. Travellers unacquainted with the peculiar circumstances of these regions are apt to neglect, or are uninformed what proper precautions to take to provide for their own safety. Man is so constituted as to compel him, in order to enjoy a healthy state of body or mind, to sleep one-third his time; and any circumstance that deranges this natural course for any length of time, superinduces pain and disease. We are persuaded that no little of the ordinary mortality prevalent upon the banks of the Mississippi and its confluent streams, arises from undue exposure to night dews and want of rest. Regimen must be left to the habits, temperament, and pursuits of the individual; no advice from another, or even rules adopted personally, can be undeviatingly pursued.

Perplexity of mind often leads to disease. We have been forced to witness some fatal instances where death could be traced from disappointed hopes. In no country have so many instances of those unfounded inflations of mind been exhibited, as in the countries we have reviewed in this treatise. As every extravagance of expectation has been fostered, the chagrin that follows failure must be in proportion to the warmth [with] which hopes of success have been cherished.

Circumstances of bitter regret sometimes happen where the sufferer has been guilty of no other fault than credulity. Land purchases are abundant, where the purchasers struggled for life against the effects of one ruinous step. The causes are numerous why emigrants, particularly Europeans, ought to proceed with the utmost caution in the purchase of landed property. If the purchases are made from the United States' government, no apprehension need be indulged respecting title; but great care should be used in choosing the spot. The advice of persons long resident ought to be taken as it respects advantages of commerce, agriculture, health, and other local conveniences.

If the purchase is made from private persons, too much care cannot be used in conveyance. In the state of Louisiana and Missouri territory, landed estate is tacitly mortgaged for its own price,—consequently, it becomes the imperative duty of a purchaser to ascertain that the payments have been made upon former sales, and that the chain of title is clear from the first grantee to the seller.

The most radical fault committed by emigrants respecting land, is, the purchase of too much, and the investment of capital in that manner, which could be much more beneficially employed upon the complete cultivation of a lesser quantity. The probable rise in the price of land is no excuse for this error. Where one man has gained by the augmentation in value of land, fifty have become rich by its fruits. The grasping at wide spaces of soil is a natural consequence of the great expanse upon which men exist in new settlements. The accumulation of land assumes the madness of avarice. Land is possessed not from any prospect of cultivation, but from vanity.

So prevalent is the foregoing propensity in the western states, that many persons are ingulphed unwarily, who would, upon mature reflection, severely condemn their own inadvertence. It may not be thought probable, but is nevertheless a fact, that within the last twenty years no subject has been more productive of ruin, to the people of the western states, than indiscreet land purchases.

The farmer, who with a moderate capital and a family, ought to prefer a small, fertile and well situated tract as his place of beginning. His surplus ought to be appropriated to improvement, and will if judiciously applied produce more and in a shorter time than if vested in superfluous landed estate.

To men who remove into the western or southern estates with money, this is a rock of temptation upon which they are very liable to be dashed. So many have involved themselves by purchasing land, that every lure is laid before the monied emigrant to induce him to relieve, by his purse, embarrassments created by the very folly he is now solicited to commit.

It might be expected that something ought to be addressed to professional men. There is, however, but one observation that can be made as respects either of the learned professions, that they have the same chances of success as other classes in society, if removing to the westward. The same perseverance, attention to their respective duties, and superiority of talent, which ensures superiority in other pursuits, will produce the same effect with the lawyer or the physician. We can only say, we have never known an individual fail, from Pittsburg to New-Orleans, in either of the two latter professions, who deserved to prosper.

To merchants nothing need be addressed. The nature of mercantile transactions are nearly the same in all places.

In enumerating the list of authors who have written upon any part of the Ohio and Mississippi valleys, it may excite some surprise to find the list so small; but it would have been difficult to enlarge it, without including names that, to speak charitably, would convey no useful information. (pp. 293–98)

# A Tour from the City
# of New-York, to Detroit

The text comes from the 1819 Kirk and Mercein edition, as reprinted in 1962 by Quadrangle Books, Chicago. Darby's *Tour* consists of sixteen letters composed en route from New York City to Detroit between May 2 and September 22, 1818. He embarked on the journey to participate in a government survey of the St. Lawrence River, but he found the work tedious and so set out for the Michigan Territory, describing the sights along the way for an unidentified friend in New York. Upon his return to the city, the author augmented his travel correspondence with notes from his 1816 journey up the Hudson River; dozens of lengthy footnotes; a letter to Charles G. Haines, Secretary of the New-York Corresponding Association for the Promotion of Internal Improvements; and an appendix containing excerpts from another geographical study, more correspondence, and additional remarks on internal improvement. Darby also prepared two maps for the volume: one of the American West (as he knew it), comprehending the region from the Hudson River west to the Missouri Territory and from the Great Lakes south to Kentucky; the other showing the Straits of Niagara. Though beset by the occasional awkwardness typical of his style, the *Tour* contains some of Darby's most charming sketches of the natural landscape and settlements of the West.

## from *Preface*

I cannot conceive the satisfaction it can give, to a generous and feeling heart to trace the last fragments of a ruined city, or behold reduced to desolation, fields that once waved in golden harvest. To the eye of reason and philosophy, a review may be desirable of the revolutions of human society in all the various stages from the savage horde to the most refined civilization; but to me, it would yield more pain than gratification, to behold Rome, Athens, or Jerusalem, in dust and ashes. The reminiscence that should recall former greatness, that would raise in imagination from the tomb the Pericles, Euripides, Maccabees, the Scipios or the Caesars, would excite, rather a tear of bitter regret, than a pleasing sentiment of poetic enthusiasm, on glancing over the ocean of past time. I would rather indulge my fancy in following the future progress, than in surveying the wreck of human happiness; I would rather see one flourishing village rising from the American wilderness, than behold the ruins of Balbec, Palmyra, and Persepolis.

Like Chauteaubriand, I have often *reposed in the woods and plains of North America*, in the silence of night, under the glances of the swan of Leda, the gleams of Sirius, or the beams of the pale moon playing amid the leaves of the forest, or exhibiting the fairy picture of the distant prairie. I have thus often in the awful solitude of the cane brake, or the cedar groves, contemplated the rapid march of active industry; I have fancied the rise of towns and villages, the clearing of fields, the creation of rich harvests, of orchards, meadows, and pastures. I have beheld the deep gloom around me dispelled, the majestic but dreary forest disappeared, the savage was turned into civilized man; schools, colleges, churches, and legislative halls arose. The river, upon whose banks now grew the tangled vine, and in whose waters the loathsome alligator floated, became covered with barks loaded with the produce of its shores; I heard the songs of joy and gladness; I beheld fair science shed her smiles upon a happy and enlightened people; I beheld the heavenly form of religion, clothed in the simple garb of love and truth, teaching the precepts of present and everlasting peace; I saw liberty and law interposing between the shafts of oppression and the bosom of innocence;—and I saw the stern brow of justice bedewed with a tear over the chastised victim.

Many were the long and tedious hours I have thus beguiled, when no sound interrupted my chain of reflection, except the sighing of the nightly breeze, and I have enjoyed a pleasure greater than man ever felt amongst "*broken columns and* disjointed arcades." I have seen on an immense surface, these warm anticipations realized. In west Virginia, in west Pennsylvania, in Kentucky, Ohio, Indiana, Illinois, Tennessee, Missouri, Mississippi, Louisiana, and Alabama; in west New-York, Michigan, and in Canada, I have for thirty-five years, been a witness to the change of a wilderness to a cultivated garden. I have roamed in forests, and upon the same ground now stand legislative halls, and, temples of religion. New states have risen, and are daily rising upon this once dreary waste. I am willing to leave the man unenvied to his enjoyments, who would prefer the barbaric picture now presented by Greece, Asia Minor, Syria, and Palestine, to the glowing canvas whose tints are daily becoming richer and stronger, upon the rivers and hills of North America. I would rather read the immortal works of Homer, Thucidydes, or Demosthenes, upon the banks of the Ohio or St. Lawrence, than search the deserted tombs of those mighty geniuses, in their now desolate native land. These men have left their bones to oblivion, their works they have bequeathed to the human race. Amid the thousand objects that are constantly before the mental eye, in this new moral creation, none is more wonderful or more alluring than the existence of more than a thousand seminaries of education, where less than thirty years past, stood no mansion of civilized man. (pp. iv–vi)

## from *Letter I*

Albany, May 3d, 1818.

Dear Sir,

Amid the violence of wind and rain, I arrived in this city at 5 o'clock this afternoon. Though spring has made some advances near New-York, here the face of nature is marked with all the bleakness of winter, except snow. At this season, no scenery can exhibit a more dreary aspect than that of the Hudson; naked rocks or precipices, with a few leafless forest trees, are the only objects that in many places meet the eye of the voyager in passing many miles upon this truly singular river. While the cold damp wind sweeps along the current, the view of the distant farm houses have a solitary and even gloomy appearance.

Perhaps in no equal distance on earth, is the contrast between the smiles of summer and the frowns of winter, so strong as upon the Hudson banks between New-York and Albany. I travelled upon both shores of the Hudson river in the summer of 1816, and visited most places of note on or near its margin.

I had then occasion to make a remark I have since found just; that the arrangements of the Steam-Boats, deprive passengers of the view of much of the richest scenery of this interesting region. The passage of the river, through the Fishkill mountains, is indeed one of the finest land-scapes in North America, and yet is seen but by very few of those who traverse through its sublime portals, and who travel expressly for the purpose of beholding nature in her most attractive garb. In the first in-stance, travelling by a land conveyance and by slow stages, I had the ad-vantage of beholding the various parts rather more in detail, than I could have, had I passed by the ordinary means of the Steam-Boat. As you have imposed upon me the task of relating what I have seen or thought, and as you have had the kindness to express more estimation for the mat-ter than the manner, I will give a detail of my notes, during my first voy-age up the Hudson.

I left the city of New-York, on the afternoon of August 20th, 1816; the weather was extremely boisterous for the season; a strong north wind impeded the progress of the Steam-Boat, and as usual, I passed the High-lands in the night, and landed about midnight at Newburg. (pp. 9–10)

*     *     *     *

During the afternoon of Aug. 21st, I crossed the river from Newburg to Fishkill landing, enjoyed in the traverse, the changing view of the nar-rows, and after landing, turned and beheld the two villages of Newburg and New-Windsor hanging upon the western slope of the opposite shore. I had here again another opportunity of admiring the ever varying scenery of this truly delightful neighborhood. Often as I have beheld with a sensation of real pleasure, the setting of an unclouded sun, never before (or since) did I see that luminary take his nightly leave of man, with more serene majesty, or amid so many objects to heighten the beauty of the scene. Seated upon an elevated bank, in a grove composed of spruce and cedar, I watched the departure of the king of day; the slow and silent advance of darkness, at length shrouded in gloom a picture, whose teints can only be forgotten when my bosom ceases to beat.

Environed by the massy and sublime monuments reared by the hand of nature, and enjoying the softened beauty of such an evening, I could not repress a retrospection upon the march of time; I could not avoid re-flecting that an epoch did exist, when the delightful valley in which I then sat was an expanse of water; that the winding and contracting gorge, through which the Hudson now flows, did not exist, or was the scene of another Niagara; I beheld the lake disappear, the roar of the cataract had ceased, the enormous rocky barriers had yielded to the impetuous flood. The river now glides smooth and tranquil, in its passage through this glen, dark and deep. The war of elements have subsided. The mountains have apparently separated, and given the waters free egress to the ocean.

In order to have ample means of reviewing this region, to as much advantage as possible, I hired a man to convey me in a sail boat, from Fishkill landing to West-Point; and on the morning of the 22d, passed the Narrows with a light wind. A slight mist floated over the highest peaks of the mountains, but below the air was clear and pleasant. Approaching the most confined part of the passage, the vast granitic ledges seemed to raise their frowning projections to the clouds, the trees upon their summits appear like shrubs. In the intervening vales or rather ravines, the fisherman and woodcutter have reared their huts; the curling smoke is seen issuing from cabins embosomed amid these rugged rocks.

West-Point presented its structures perched upon a small cape of level land, but every where surrounded by masses that seemed to mock time itself.

I landed, and rose the winding path that led to this ever-memorable spot; a place that was the scene of some of the most remarkable events of our unequalled revolution. It was here that Arnold's treachery was met by the stern virtue of Washington; it was near this place that Andre expiated his folly with his life, and gained an immortal name by an ignominious death.

West-Point presents but little that can interest the traveller, except it be the noble scenery of its neighborhood, and events of historical reminiscence. The barracks of the officers and cadets, with a few scattering houses belonging to individuals, are all the artificial improvements worth notice at this establishment. The bank is high, and very abrupt from the surface of the water in the river, to the level of the plain upon which the barracks and houses are built.

With considerable fatigue, I scrambled up the mountain to the ruins of Fort Putnam. Silence and delapidation now reign over this once important Fortress. It would be difficult to conceive of a more impregnable position. Seated upon an elevated mass of granite, the Fort occupied almost the entire surface upon which a human foot could be set. A very steep ascent, of more than 500 perpendicular feet, leads from the plain of West-Point to the site of the Fort, and a deep rock bound valley, separates it from the general mass of the adjacent mountains. A cistern had been hewed out of the solid granite, which was full of water when I visited the spot. Cannon placed upon the walls of this Fort, could rake the entire surface of West-Point; but I could not perceive any serious opposition it could have presented to the passage of ships of war, ascending or descending the Hudson river.

The landscape from the ruined battlements of Fort Putnam, is very interesting. The Fishkill mountains seen from this place, have a much more naked and rude aspect, than from either Newburg or Fishkill landing. Except upon the opposite shore in Putnam county, but very little human culture enlivens the view. West-Point has itself a solitary appearance, and to the west, nought is seen but woods, and mountains, in their primitive wildness.

If seclusion from the busy haunts of men, can be of any benefit to the students at West-Point, they enjoy this advantage in its fullest extent. Isolated upon the confined cape, from which the name of the place is derived, the river on one side and towering mountains on the other, an unbroken silence reigns around this seminary. Looking down from the broken walls of Fort Putnam, Dr. Johnson's Rasselas, came strong to recollection. I could not avoid recalling to imaginary life, the men who once acted on this little but remarkable theatre. I felt a sentiment of awe, amid this now lonely waste, on recalling to mind that here once depended the fate of a new born nation. Even the fallen fragments of stone which once composed part of its buttresses, inspired me with a feeling of respect. Washington, Greene, Putnam, Andre, and Arnold, are no more; their names have now taken their respective stations in history. The opinion of mankind is formed upon the merits of the three former, and the shame of the two latter. It is now as far beyond the reach of calumny, to tarnish the unfading renown of a Washington, a Greene, or a Putnam, as it would be for the human hand to level to common earth the enormous masses of the Fishkill mountains.

With slow steps I descended from the grey remains of this venerable pile, and cast a frequent and repeated retiring look towards its mouldering turrets. The shades of evening were setting in, the darkened sides of the distant mountains, seemed to mark a sympathetic gloom with that which hung over the deserted Fortress. The busy hum of the students in their evening walks, produced an interesting contrast with the repose in which rested the surrounding scenery. Such were the events, and the reflection of my day's visit to the West-Point. (pp. 11–15)

## from *Letter XIII (Buffalo, July 31, 1818)*

July 29th I visited Black Rock. This is a small but apparently a thriving village, two miles north of, and built upon the same plain with Buffalo. Here the banks of the Niagara river or strait, present a very exact resemblance with those of the St. Lawrence, from Brockville to Hamilton. Rising by gentle acclivity from the water; both sides of the river being cultivated afford a fine prospect, though from its longer settlement, the Canada shore is much more improved than that of New-York. Unless in a cataract, I never before witnessed so large a mass of water flowing with such prodigious rapidity. The bottom of the river is composed of smooth rock, over which the water glides. If the stream flowed over broken masses of stone it would be impassable.

After viewing Black Rock I took advantage of a boat going down, and hasted towards one of the great objects of my journey, the Falls of Niagara. The day was intolerably warm, with scarce an air of wind to move a leaf. I found the river much more winding than I had expected from the maps I had seen. Our boat followed the west channel, leaving Grand island to the east. Passing this island I was struck with its remarkable resemblance to many of the St. Lawrence islands, having a similar swell rising from the water. Some new openings are now making, but the greatest part of its surface is yet forest. I had no means of examining the timber, but at a distance the trees had a similar mixture with the opposite shores, hemlock, sugar maple, elm, oak, and linden.

Tonnewanta and Ellicott's creek enter the east channel of Niagara strait very nearly opposite to the middle of Grand island. Extensive marshes and swamps skirt the Tonnewanta from its mouth, for more than twenty miles upwards. This creek or rather river, rises in the town of Orangeville, in the south side of Genesee county, interlocking with the sources of Cataraugus and Buffalo creeks, and with some streams which enter the west branch of Genesee river. From its source the Tonnewanta crosses in a northern direction Orangeville, Attica, and Alexander townships, reaches Batavia after flowing about twenty-five miles; it thence gradually curves to the north-west, west, and south-west by west, falls into Niagara river forty miles from Batavia, having an entire course of sixty five miles. This stream has now become an object of interest, from the circumstance of its bed being for some distance intended as the route of the Grand Canal; the land contiguous to the lower part of its course from Batavia, is, as I have already observed, subject in many places, to submersion by water. It is navigable for boats upwards of twenty miles from its mouth. Between the mouth of the Tonnewanta and old Fort

Schlosser, the marshes in some places border the strait; and what is remarkable, the Chippewa river entering the Canada side a short distance
above the falls, exhibits in some measure, similar phenomena with the
Tonnewanta. Seen from the strait below the lower extremity of Grand
isle, the whole adjacent country appears almost level, no elevation being
visible that materially breaks the monotony of the landscape. The strait
here turns nearly abruptly to the west, and first exposes to view the cloud
that constantly rises from the cataract. Nothing is seen, however, that anticipates in any manner the sublime and awful scene below; even the
rapid current that sweeps past Black Rock, is now tranquillized; the strait
is here nearly as still as a lake on the U.S. shore, and flows gently on that
of Canada. Navy island is a small extent of land lying in the Canada
channel, at the lower extremity of Grand island, below which commences the rapids that precede the cataract of Niagara. I passed between
Navy and Grand islands, and landed near old Fort Schlosser, and walked
down the shore to Whitney's, opposite the *falls*; it was near sun-set, silence began to reign over the face of nature. Slowly and at intervals I
heard the deep, long, and awful roar of the cataract; my mind which for
years had dwelt with anticipation upon this greatest of the world's traits,
approached the scene with fearful solicitude. I beheld the permanent objects, the trees, the rocks; and I beheld also the passing clouds, that mo-
mentarily flitted over the most interesting picture that nature ever painted
and exposed to the admiration of intelligent beings, with more than my
common forbearance, I concluded to behold amid the beams of a rising
sun the greatest object ever presented to human view. But whilst the stars
of the night gleamed through the misty atmosphere of this apparently
fairy land, I walked forth to the margin of the cataract, and in fancy conceived the beauties, the horrors, and the wonders the coming morn
would produce. That morn opened, (July 30th) it was clear and serene; I
hasted to the verge of the cataract; I expected much, and was not disappointed. The point of land above A. is a thick wood standing upon a
sloping bank. The noise of the cataract is heard, but its features unseen,
until the observer advances to the verge of the fall; it is then seen so
obliquely as to destroy its best effect. Defective, however, as was this
perspective of Niagara, it presented beauties infinitely transcending any I
had ever seen before. I stood upon the very slope over which the torrent
rushed, and for many minutes forgot every other object except the unde-
scribable scene before me; but when the fervor of imagination had in
some measure subsided, I beheld under my feet, carved on the smooth
rock G. D. C.; W. P. and J. B. and many other initials of friends that had
visited this incomparable spot, and left these memoria, that friends only

could understand. On beholding these recollections of home, you will forgive me when I acknowledge having dropt upon their traces tears, that were rapidly swallowed in the vortex of Niagara. The beams of morning came, and glanced upon the curling volumes that rose from the abyss beneath; my eye searched the bottom of this awful gulf, and found in its bosom darkness, gloom, and indescribable tumult. My reflections dwelt upon this never ending conflict, this eternal march of the elements, and my very soul shrunk back upon itself. The shelving rock on which I stood trembling under my feet, and the irresistible flood before me seemed to present the pictured image of evanescence. The rock was yielding piecemeal to ruin, fragment after fragment was borne into the terrible chasm beneath; and the very stream that hurried these broken morsels to destruction, was itself a monument of changing power.

I retraced my steps to Col. Whitney's, and after breakfast returned, and descending the almost perpendicular bank of rocks, found myself under the tremendous FALL OF WATER, that even in description has excited the admiration of cultivated man! I crossed the Niagara strait about 250 yards below the *chute*. The river was in some measure ruffled by the conflict it had sustained above, but no danger approached the passenger. Perpendicular walls of rock rose on both sides, to the appalling elevation of between three and four hundred feet. The trees which crowned the upper verge of this abyss appeared like shrubs. I was drenched to the skin by the spray of the cataract; but the sublime scene towering over my head, was too impressive to permit much reflection upon a momentary inconvenience. The river below the fall flows with considerable rapidity, but with less velocity or turbulence than I had been induced to expect. The opposing banks are perfectly similar, both being perpendicular about half the descent; below which enormous walls extend slopes, composed of the broken fragments that have been torn from their original position by the torrents from above. Most maps of Niagara are very defective, the river being represented too straight. The best delineation of this phenomenon which I have seen is contained in the map of Niagara river, published with Gen. Wilkinson's Memoirs. In that draft, the river above the falls is represented, as it is in fact, flowing almost westward. Below the *chute* the stream flows abruptly to the north-east, which course it pursues more than a mile, from whence it again resumes a northern direction, which, with some partial bends, it continues to the place of its final exit in lake Ontario. (pp. 158–62)

*     *     *

No adequate idea can be formed from description of this wonder of interior North America. Its pitch in fact, its width, velocity, and consequent mass, can be estimated with considerable accuracy; but the effect upon the mind can only be produced from actual view. If the massy walls of rock, and the rapids above are excepted, there is nothing near Niagara that is striking in the scenery. It is left alone in simple and sublime dignity to strike the soul with a sensation that loss of life or sense alone can obliterate, but the nature of which no language can convey. If towering mountains and craggy rocks surrounded Niagara, I cannot but believe that much of its fine effect would be lost; as it exists it is an image whose whole contour is at once seen, and the recollection unbroken by extraneous objects; even sound is subservient to the impression made upon the heart, [and] none is heard except the eternal roar of the cataract. I would have been rejoiced to have seen this place in a tempest. The whole time I was there, the weather, though warm, was otherwise serene and pleasant. Amid the howling of the black north-west wind Niagara must have something of more than common interest. I am inclined nevertheless to believe, that winter alone can give all its most appropriate attendant imagery to the falls. But at all times, at all seasons, and I might say by all minds, will this matchless picture be viewed with wonder and delight, and remembered with feelings of pleasure. (pp. 163–65)

*     *     *

It is when standing upon the brow of these heights, that the fact becomes demonstrative that here once dashed Niagara, mingling his foaming surge with the wave of Ontario. The rocky bed has yielded to the ever rolling waters, and the cataract has retired to the deep and distant dell where it now repeats the thunders of ages, and continues its slow but certain march to Erie. Time was when Niagara did not exist, and time will come when it will cease to be! But to these mighty revolutions, the change of empire is as the bursting bubble on the rippling pool, to the overwhelming volume that rolls down the steep of Niagara itself. Since this cataract fell where Queenston now stands, have risen and fallen Assyria, and Persia; Macedonia, and Rome; the flood of northern barbarians issued forth from their native woods, and in the storm of savage fury profaned the tombs of the Fabii, and the Scipio's, and in the march of time the polished sons of those mail clad warriors, now seek with religious veneration the fragments of the statues that their fathers broke; and whilst this moral stream was flowing through the wide expanse of ages, has the Niagara continued its unceasing course. Roused from the sleep of a *thousand years*, the energies of the human mind sought another world, and found America; and amid this new creation found Niagara. During the change of nations, religion and language, this vast, this fearful cataract unceasingly pursued and pursues its slow and toilsome way. (pp. 168–69)

# Bibliography

## MANUSCRIPT COLLECTIONS

Draper Manuscript Collection, Vol. 8 NN, State Historical Society of Wisconsin, Madison, Wis.

Peter Force Papers, Library of Congress, Washington, D.C.

Simon Gratz Collection, Historical Society of Pennsylvania, Philadelphia, Pa.

Andrew Jackson Papers, Library of Congress, Washington, D.C.

Miscellaneous Collection, Louisiana State University Archives, Baton Rouge, La.

Miscellaneous Collection, New-York Historical Society, New York, N.Y.

## MAGAZINES AND NEWSPAPERS

*American Review* (later, *American Whig Review*), 1845–51
*Casket*, 1829–38
*Darby's Monthly Geographical, Historical, and Statistical Repository*, 1824
*Gazette and Universal Daily Advertiser* (Philadelphia), 1833–34
*Monthly Journal of Agriculture*, 1845–46
*National Gazette and Literary Register* (Philadelphia), 1820–22
*National Intelligencer* (Washington, D.C.), 1830–54
*Niles' Weekly Register*, 1815–50

*Reporter* (Washington, Pa.), 1845–46
*Saturday Evening Post*, 1829–38

## LITERARY, HISTORICAL, AND BIOGRAPHICAL SOURCES

Albach, James R. *Annals of the West*. Pittsburgh: W. S. Haven, 1858.

Albert, George Dallas. "The Frontier Forts of Western Pennsylvania." In *The Frontier Forts of Pennsylvania*. By the Commission to Locate the Site of the Frontier Forts of Pa. Harrisburg: Clarence M. Busch, 1896. Vol. II.

Alden, John R. *Rise of the American Republic*. New York: Harper and Row, 1963.

Ames, William E. *A History of the National Intelligencer*. Chapel Hill: University of North Carolina Press, 1972.

Atkeson, Mary Meek. *A Study of the Local Literature of the Upper Ohio Valley*. Ohio State University Bulletin: Contributions in English, Vol. XXVI. Columbus: Ohio State University, 1921.

Bakeless, John. *Daniel Boone*. 1939; rpt. Harrisburg: Stackpole Co., 1965.

Billington, Ray Allen. *Westward Expansion: A History of the American Frontier*. New York: Macmillan Co., 1949.

Brown, Wilburt S. *The Amphibious Campaign for West Florida and Louisiana, 1814–1815*. University, Alabama: University of Alabama Press, 1969.

Butterfield, Consul W. *An Historical Account of the Expedition Against Sandusky*. Cincinnati: Robert Clarke, 1873.

Darby, William, ed. *Brookes' Universal Gazetteer*, 3rd American ed. Philadelphia: Bennett and Walton, 1823.

———. *The Emigrant's Guide to the Western and Southwestern States and Territories*. New York: Kirk and Mercein, 1818.

———. *A Geographical Description of the State of Louisiana*. Philadelphia: John Melish, 1816.

———. *Lectures on the Discovery of America, and Colonization of North America by the English*. Baltimore: Plaskitt and Co., 1828.

———. *Memoir on the Geography, and Natural and Civil History of Florida*. Philadelphia: T. H. Palmer, 1821.

———. *Mnemonika; or, The Tablet of Memory*. Baltimore: E. J. Coale, 1829.

————— and Theodore Dwight, Jr. *A New Gazetteer of the United States of America*. New York: Edward Hopkins, 1832.

—————. *The Northern Nations of Europe, Russia and Poland*. Chillicothe, Ohio: S. W. Ely, 1841.

—————. *Remarks on the Tendency of the Constitution of the United States to Give Legislative Control to the President*. Washington: n.p., 1842.

—————. *A Tour from the City of New-York, to Detroit*. New York: Kirk and Mercein, 1819.

—————, ed. *The United States Reader, or Juvenile Instructor*, 2nd ed. rev. Baltimore: Plaskitt and Co., 1830.

—————. *View of the United States, Historical, Geographical, and Statistical*. Philadelphia: H. S. Tanner, 1828.

De Hass, Wills. *History of the Early Settlement and Indian Wars of Western Virginia*. 1851; rpt. Parsons, W. Va.: McClain Printing, 1961.

Doddridge, Joseph. *Notes on the Settlement and Indian Wars of the Western Parts of Virginia and Pennsylvania*. 1824; rpt. Parsons, W. Va.: McClain Printing, 1960.

Downes, Randolph C. *Council Fires on the Upper Ohio*. Pittsburgh: University of Pittsburgh Press, 1940.

Egle, William H. *The Dixons of Dixon's Ford*. Harrisburg: Dauphin County Historical Society, 1878.

—————, ed. "Autobiographical Letter of William Darby." In *Notes and Queries Historical and Genealogical*. Harrisburg: Harrisburg Publishing Co., 1894. Vol. I, pp. 33–41.

Fiedler, Leslie. *Love and Death in the American Novel*. 1960; rpt. New York: Dell Publishing, 1966.

Flint, Timothy. *A Condensed Geography and History of the Western States, or the Mississippi Valley*. 2 vols. Cincinnati: E. H. Flint (Vol. I); William M. Farnsworth (Vol. II), 1828.

Forrest, Earle R. *History of Washington County, Pennsylvania*. Chicago: S. J. Clarke, 1926. Vol. I.

Fussell, Edwin. *Frontier: American Literature and the American West*. Princeton: Princeton University Press, 1966.

Kennedy, J. Gerald. "Glimpses of the 'Heroic Age': William Darby's Letters to Lyman C. Draper." *Western Pennsylvania Historical Magazine*, LXIII (January, 1980), 37–48.

Lockett, Samuel H. *Louisiana As It Is: A Geographical and Topographical Description of the State*. ed. Lauren C. Post. Baton Rouge: Louisiana State University Press, 1969.

Loudon, Archibald. *A Selection of . . . Narratives, of Outrages, Committed by the Indians, in their Wars, with the White People.* 2 vols. 1808; rpt. New York: Arno Press, 1971.

McLemore, Richard Aubrey, ed. *A History of Mississippi.* Hattiesburg: University and College Press of Mississippi, 1973. Vol. I.

Miller, Perry. *The Raven and the Whale.* New York: Harcourt Brace, 1956.

Mott, Frank Luther. *A History of American Magazines.* Cambridge: Harvard University Press, 1939. Vol. I.

O'Rielly, Henry. "Pioneer Geographic Researches." *Dawson's Historical Magazine, XIII* (October, 1867)`, 223–26.

Ross, Frank Edward. "William Darby." In *Dictionary of American Biography*, ed. Allen Johnson and Dumas Malone. New York: Charles Scribner's Sons, 1943. Vol. V, p. 73.

Rusk, Ralph Leslie. *The Literature of the Middle Western Frontier.* 2 vols. New York: Columbia University Press, 1926.

Schlesinger, Arthur M., Jr. *The Age of Jackson.* Boston: Little, Brown, and Co., 1945.

Sipe, C. Hale. *The Indian Wars of Pennsylvania.* Harrisburg: Telegraph Press, 1929.

Slotkin, Richard. *Regeneration Through Violence: The Mythology of the American Frontier, 1600–1860.* Middletown, Conn.: Wesleyan University Press, 1973.

Smith, Henry Nash. *Virgin Land: The American West as Symbol and Myth.* New York: Vintage Books, 1950.

Smith, Joseph. *History of Jefferson College.* Pittsburgh: J. T. Shryock, 1857.

Spengemann, William C. *The Adventurous Muse: The Poetics of American Fiction, 1789–1900.* New Haven: Yale University Press, 1977.

Turner, Frederick Jackson. *The Frontier in American History.* 1920; rpt. New York: Holt, Rinehart and Winston, 1962.

Vail, R. W. G. *Knickerbocker Birthday: A Sesqui-Centennial History of the New-York Historical Society.* New York: New-York Historical Society, 1954.

Withers, Alexander Scott. *Chronicles of Border Warfare.* 1831; rpt. Parsons, W. Va.: McClain Printing, 1961.

# Index